RAM DASS

For those who have followed Richard Alpert's journey from a professorship at Harvard through experimentation with LSD to the Himalayas in India, where he was christened Ram Dass (Servant of God) by his spiritual teacher, this book speaks in intimate and illuminating terms of the paths taken and the paths yet to be traveled. A time will come, says the author, when we will be able to gather and sit in silence, not in expectation, but in fulfillment. *Grist For the Mill* brings us closer to that time, through its cosmic mapping of the evolutionary cycle and levels of reality.

"I really love *Grist* for it's the closest thing vibrationally to a sequel to *Be Here Now*."

—Ram Dass

Bantam Books by Ram Dass

GRIST FOR THE MILL
JOURNEY OF AWAKENING

GRIST
FOR THE
MILL

RAM DASS
with Stephen Levine

BANTAM BOOKS · TORONTO · NEW YORK · LONDON

🐤

GRIST FOR THE MILL
*A Bantam Book / published by arrangement with
Unity Press*

PRINTING HISTORY
Unity Press edition published May 1977
2nd printing____September 1977   3rd printing_____July 1978
Parts of this book were first published in Inside/Out
and The Journal of Transpersonal Psychology
Bantam edition / April 1979

ISBN 0-553-12457-9

Published simultaneously in the United States and Canada

*Bantam Books are published by Bantam Books, Inc. Its trade-
mark, consisting of the words "Bantam Books" and the por-
trayal of a bantam, is Registered in U.S. Patent and Trademark
Office and in other countries. Marca Registrada. Bantam
Books, Inc., 666 Fifth Avenue, New York, New York 10019.*

PRINTED IN THE UNITED STATES OF AMERICA

Dedication

The Dharma belongs to no one. Teachings about the Dharma come now through one person and now through another. What is contained in this book certainly did not originate with me. It is part of a river which flows through me from Guru, teachers, parents, past incarnations and life experiences.

As I read this manuscript I can feel in the turn of a phrase or an image the intimate presence of my Guru and one or another of my teachers. Their very real contributions to this book are warmly and gratefully acknowledged. May this book serve as an expression of appreciation for their teachings.

My thanks also to Stephen Levine, co-author, whose sensitive poetic collaboration made the words come to light as I heard them but could not quite speak them.

Shanti,
Ram Dass

New York City 1976

Contents

Collaborator's Note

The space from which these understandings come has no body, no mouth, it cannot speak. To be communicated these insights had to cross the wild river of accumulated personality, acculturation, interpretation, opinion, and preference to enter into the limitations of language. They are offered as a near-translation of the experience of things as they really are. These teachings were originally offered in the direct, charismatic, air medium of the oral tradition before once again being translated and further grounded into the powerful earth medium of the written word, print on paper, book form.

The transmission from form to form continued without the self-conscious "presence" of an editor, but instead flowed from the experience out of which these teachings had originated. The continuity was grace elicited from the fullness of each moment as it manifested before us as this book. The collaboration occurred on a plane where the beings collaborating were no one in particular, so there was little to impede or diffuse the natural intensity of the light.

As this oral tradition translated itself into the written word we decided not to italicize, specialize, the Sanskrit-derived terms such as: sadhana, spiritual practice; karma, the actions of life which breed further reaction; samadhi, deeply concentrated states; Guru, the teacher, the teaching—because these concepts should not be something different, or "other," but should be allowed to enter into the marrow of the language. So too dharma, as natural karmic duty, appropriate action for this incarnation, is not capitalized for it is "nothing special;" while Dharma, the truth, Natural Law, the Tao, God's will, is capitalized to demonstrate its implacability.

Interwoven from lectures, retreats, articles and interviews given during 1974–1976 in Philadelphia, Washington, Lincoln, Seattle, Los Angeles, Boston, Portland, San Francisco, Santa Cruz, Kansas City and Aspen, these words are offered as the gift of the Dharma which is ever and always present to each of us, in each of us.

Let it Shine,
Stephen Levine

Santa Cruz 1976

Grist for the Mill

In India when we meet and part we often say, "Namasté," which means I honor the place in you where the entire universe resides, I honor the place in you of love, of light, of truth, of peace. I honor the place within you where if you are in that place in you and I am in that place in me, there is only one of us.

Namasté

The Journey

Welcome to an evening under the breast of the Divine Mother. It's so graceful to share the journey. We've been on the journey a long time together. We've gone through a lot of stages. And just as in any journey some people have dropped along the way, have had enough for this round. Others have been waiting for us to catch up. The journey passes thru the seven valleys, the seven kingdoms, the chakras, the planes of consciousness, the degrees of faith. Often we only know we've been in a certain place when we pass beyond it, because when we're in it we don't have the perspective to know because we're only being. But as the journey progresses, less and less do you need to know. When the faith is strong enough it is sufficient just to be. It's a journey towards simplicity, towards quietness, towards a kind of joy that is not in time. It's a journey out of time, leaving behind every model we have had of who we think we are. It involves a transformation of our beings so that our thinking mind becomes our servant rather than our master. It's a journey that has taken us from primary identification with our body, through identification with our psyche, on to an identification with our souls, then to an identification with God, and ultimately beyond identification.

Because many of us have traversed this path without maps, thinking that it was unique to us because of the peculiar way in which we were traveling, often there has been a lot of confusion. Imagining the end was reached when it was merely the first mountain

1

peak which yet hid all of the higher mountains in the distance. Many of us got enamoured because these experiences along the way were so intense that we couldn't imagine anything beyond them. Isn't it a wonderful journey that at every stage you can't imagine anything beyond it? Every point you reach is so much beyond anything up until then that your perception is full and you can't see anything else but the experience itself.

For the first few stages you really think that you planned the trip, packed the provisions, set out yourself, and are the master of your domain. It's only after a few valleys and mountains some ways along that you begin to realize that there are silent guides, that what has seemed random and chaotic might actually have a pattern. It's very hard for a being who is totally attached and identified with his intellect to imagine that the universe could be so perfectly designed that every act, every experience is perfectly within the lawful harmony of the universe. Including all of the paradoxes. The statement, "Not a leaf turns but that God is behind it," is just too far out to think about. But eventually we begin to recognize that the journey may be stretching out for a longer span than we thought it was going to.

We come out of a philosophical materialistic framework in which we are totally identified with our bodies and the material plane of existence, and when you're dead you're dead so get it while it's hot. And more is better and now is best, because you don't know when the curtain will come down and it will all be over. And better not to think about that curtain because it's too frightening. Where along the journey do you begin to suspect that that model of how it is is just another model? And that this lifetime is but another part of a long, long journey? In the Buddhist teachings there is an analogy of how long we've been doing this. The image is that of a solid granite mountain six miles long, six miles wide, six miles high. Every hundred years a bird flies by the mountain with a silk scarf in its beak and runs the scarf over the mountain. In the length of time it takes for the silk scarf to wear away the mountain, that's how long we've been doing it. Round after round after round. It puts a different time perspective on this one, doesn't it? Not all of those rounds on this plane, not all of those rounds in human form. But all of those rounds are a part of a journey that has direction.

Sooner or later the realization comes that nothing you can think of is going to do it. Nothing you experience is it. Because your mind thinks of things and you and the thing are separate and there is a little veil, like a trillionth of a second that exists between you and the thing you're thinking of. And when you sense something or collect an experience, there's the distinction between the experience and the experiencer and that's a very thin veil. It doesn't matter how thin it is, it's like steel. It always separates you from where it's happening.

When at last the despair is deep enough you cry out. You cry inwardly or outwardly, "Get me out of this! I want to get out! I give up. I don't know. I surrender." At that moment when the despair is genuine enough the veil separates a bit. I'm not talking about the wanting to want to give up. I'm not even talking about the wanting to give up. I'm talking about the giving up. The problem is most of us say, "I don't think my thoughts are going to do it so I'm now open to new possibilities. I'll read Ram Dass' book but I'm going to sit and judge it." Forget it. Because the judge has designed the game so that the judge won't have to change, and says, "Anything that doesn't fit in with the way I thought it was I reject." We have categories for that, it's "weird" or it's "occult" or it's "far-out" or whatever you want to call it. It's a way of putting it somewhere else so it doesn't blow our scene up. That's what the judge's function is, so the scene doesn't get blown up.

When as the Third Patriarch suggests you set aside opinion and judgment because you see they're just digging you deeper into your hole, you surrender your own knowing. Now that's really hard because the whole culture is based on the worship of the golden calf of the rational mind while other levels of knowing,

3

like what we call intuition, have practically become dirty words in our culture. It's sort of sloppy, it's not tight, logical, analytic, clean. You don't sit in scientific meetings and say, "I intuit that. . . ." You say, "Out of inductive reasoning I hypothesize that we will be able to disprove the null hypothesis." That's saying the same thing but you've made believe that you're doing it analytically and logically. Some of you I am sure recognize that game. When Einstein said, "I did not arrive at my understanding of the fundamental laws of the universe through my rational mind," many of his colleagues thought him quite eccentric—because the rational mind has been the high priest of the society. Realize that it's merely a tiny system and that there are meta-systems and meta-meta-systems in which only when you transcend your logical analytic mind can you even enter the gate.

I remember as a social scientist I studied what was studiable. What was studiable had nothing to do with what was happening to me but it was studiable. The analogy is the drunk looking for the watch under the streetlight. Someone comes to help him look but there's no watch under the streetlight and finally he asks, "Well, exactly where did you lose it?" And he says, "I lost it up in the dark alley but there is more light out here." It is the light of the analytic mind we were using to try to find what had been lost in the distant alley.

Well, a long time ago, you were enamoured of your prehensile capabilities, the fact that your thumb and index finger could do it and no other species could. That was pretty far out, you got a lot of power. But that was nothing like all the anticipatory stuff and the remembering and all this stuff you could do with your cerebral cortex. And to think that wasn't to be the end all. It even sent people to the moon. Isn't that the penultimate? It doesn't seem to be, does it? It's interesting that people were burned at the stake for suggesting that the anthropomorphic view of the universe wasn't the final one. We all have been caught again and again in embracing the view that the physical universe is the center of it all, when in fact it turns

4

out that the physical universe is just another one. Not even necessarily the most interesting one. Isn't that damaging to our egos? And the moment when there is that little bit of giving up, whether you're blown out of your rational mind by whatever techniques you have available, or some traumatic experience happens in your life which shakes you out of it, or you have just lived long enough that you've despaired of ever getting it the way you thought you were going to. Whatever the genesis of it, at that moment you experience the presence of another set of possibilities of who you are and what it's all about. It is like that moment depicted on the ceiling of the Sistine Chapel where the hands of God and man are just about to touch, it's just at the moment when the despair is greatest, when you reach up, that the grace descends and you experience the knowledge or the insight or the remembrance that it all isn't in fact the way you thought it was. If it happens too violently you decide you've gone insane. And there are people who are all too willing to reassure you that you have, and there are places for that. Because in hunting tribes, mystics are treated as insane, they're an inconvenience because the tribe has to be kept mobile and old people and crazy people have to be put away somewhere. But if you're in a certain position at the moment of seeing through, if the view has been gentle or if you're with somebody else that knows, or if you had intellectually known but didn't believe, all of which is a karmic matter, if you had some kind of structure or support system, you say "Even though everybody else thinks I'm mad, I'm not." It's like when I was being thrown out of Harvard. There was a press conference and all the reporters and photographers were interviewing me because I was the first professor to be fired from Harvard in a very long time. They all were looking at me as the fighter who had just lost the big fight. Here I was a good boy who had built his career and finally reached Harvard and now was obviously going to disappear into ignominy. The major teaching institution had dismissed him in disgrace. They had that look on their faces you have when you're around a loser. And here I was every few

5

days taking acid with my partner, Timothy, and my friends were going into these realms beyond realms beyond realms, and I was looking at the reporters and photographers as "those poor fellows." And I looked around and saw that everybody believed in only one reality to this situation except me, and I remembered, since I'm a clinical psychologist, that that was a definition of insanity. One of me was saying, "Boy, are you crazy." And the other was saying, "Go, go, go, you're right on!"

The moment at which you look up, the moment at which you look in, starts the journey back. The journey has gone from the One into the incredible paranoid multiplicity of this high technological materialistic structure. Then when the despair is great enough, there is the turn around and you start to go back to the One. And at that moment who you are starts to change. Because up until then you have been worshipping your individual differences, "I'm more beautiful, I'm younger, I'm smarter, or I wish I were," or totally preoccupied with getting an individual difference that you could accentuate, because that's where the payoff was. So you dressed in silver sequins with golden blups and that made you special and everybody said, "You're special." But then when you look around and you get a sense of another reality, an awareness of presence, a place within you starts to draw you as inevitably and irrevocably as a flame draws a moth. For a long time, maybe many, many lifetimes you'll keep soaring in close and getting your wings singed. Now whether your wing is being singed, whether the fire purifies you or destroys you depends on who you think you are because the fire can only burn your stash of clinging. The fire doesn't burn itself. And in truth, you are the fire.

Receiving the Transmission

In a gathering such as this it is no longer sufficient to just talk about it, now we have to become it. For the transmission that you come to receive is in truth not a conceptual one. What I know, I will share with you, but what I am, you must take, and for you to take what's here you have to acknowledge who you are. For if you come with the certainty that you already know and that what you have is enough, then though you will hear the words, you will not receive the transmission. If we transported this moment to a cave in the Himalayas, that you had all spent months getting to, and you entered the cave after much purification and sat before me, you would be ready to receive the transmission. Now it is a question of whether in this form, in the place you are right now, with such easy access to these words, whether the same space can be created. For by the time you got to the Himalayan cave you would recognize that what you're seeking in the transmission has very little to do with time and space, it has very little to do with your body, with your personality. It was only when don Juan had destroyed Castaneda's personal history that the transmission could occur.

When you sit in an auditorium and there is a speaker or you read a book, you tune in a set of receiving devices, your ears, your eyes and your conceptual analytic mind. But to receive this transmission requires much more than the mind. It requires a desire in you. A desire to use this birth in order to become who in truth you are. It requires the desire that you become free of the kinds of clinging and attachment that keeps distorting and narrowing your vision. It requires that you truly desire to know what you are doing here, what your function is here on Earth. To be saying I want more than the reality that is available to me through my senses and my thinking mind requires that you take those moments every one of you have had in your life when you have been in tune with the Tao, with the harmony of the universe, with the flow, when for a moment you set aside your separateness, your self consciousness, and became part of the process, in the same way as a tree or a brook or wheat is part of the process, and bring them into the foreground at this moment and make them figure so that they stand out and put the rest of the forms of your life into the background.

You had in those few moments the answer to every question your mind could create and all the food your soul needed. The only difference between me and you is that I know that and you don't, so you come here for me to remind you who you are. What you're looking for is who is looking, it's happening all around you and you are what's happening. So that next time you sit waiting for something to begin you will realize there is nothing that needs to begin, for the beginning, the middle and the end are already who you are.

Actually the wisdom I wish to share with you has no time and no space. It is not really East as opposed to West, it is not now as opposed to then. It isn't the sole domain of any organized religion, it is that which is universal to all that lives in the true spirit. Each time I go out on tour, which is less and less often, I meet us, ready to hear more. There is a percentage of you for whom the introductory rap would be appropriate. Others of you here are ready for the intermediate course, the 101 series. A few of you would like the advanced graduate seminar because you are ready to specialize, you're prepared to make a commitment in your life. For those of you who want the advanced course, if while we are together you just sit quietly and breathe in and out through your open heart, and remain very still, you will receive the transmission you seek. Don't get lost in the words, for the words are like birds, they fly in from one horizon and fly out towards the other. For those with very active minds who'd like to know what's happening, I will provide words. And you can chew on them, collect them, write them down and save them until they turn yellow, you can go away and reflect upon them until you are ready to sit quietly and open your heart and quiet your mind, for the predicament is that the transmission you in truth yearn to receive is not one that the rational mind can fully grok, can fully grasp, can fully appreciate. All the rational mind can do is get to the point where it is pointing and it says, "It went that'a way!" But to have what you seek you have to go beyond knowing and become it. It is a peculiar predicament that the knowledge can only be known by your transforming yourself into the knowledge itself.

Rules of the Game

The simple rules of this game are honesty with yourself about where you're at, and learning to listen to hear how it is. Meditation is a way of listening more and more deeply, so you hear from a more profound space, exactly how it is. To hear how it is, you must be open to it, thus the open heart. You can take your life exactly as it is in this moment; it is a fallacy to think that you're necessarily going to get closer to God by changing the form of your life, by leaving so-and-so, or changing your job, or moving, or whatever . . . by giving up your stereo, or cutting off your hair, or growing your hair, or shaving your beard, or . . . . It isn't the form of the game, it's the nature of the being that fulfills the form. If you're a lawyer, you can continue being a lawyer, you merely use being a lawyer as a way of coming to God. It is a fallacy to think that any form of life is necessarily more spiritual than any other. Ashrams are often the heaviest, most neurotic, political settings I've ever been in. They can also be very beautiful spiritual spaces, but by definition, just because you're living in some place called an ashram or a monastery, doesn't mean you're getting closer to God.

For someone who has grown in evolution to the point that she or he understands that this precious birth is an opportunity to awaken, is an opportunity to know and perhaps to be God, all of life becomes an instrument for getting there: marriage, family, job, play, travel—all of it. You spiritualize your life. When

10

Krishna says in the *Bhagavad Gita*—Do what you do, but offer the fruit to me—you do it all in relationship to becoming enlightened, so that when someone says to you, Who are you?, the answer is not I am a lawyer or a housewife. I am a being going to God. I do law in order to provide right livelihood, to protect this temple, and fulfill my responsibilities in order to go to God. I am living with so-and-so, in such and such a situation, because that situation is the optimum for me to fulfill my karma and allow me to go to God. It's as simple as that.

You find your way through this incarnation, each of us has a different path through. No path is any better than any other path, they are just different. You must honor your own path. For some of you, you will feel like half a being until you form a connection with another half and then you will be able to go to God. Others of you will go alone on your journey to God. It's not better or worse, it's just different. If you get over the value judgments you can listen to what it is you need to do without getting caught in all of the social pressures about marriage or non-marriage. The true marriage is with God. The reason that you form a conscious marriage on the physical plane with a partner is in order to do the work of coming to God together. That is the only reason for marrying when you're conscious. The only reason. If you are marrying for economics, if you're marrying for passion, if you're marrying for romantic love, if you're marrying for convenience, if you're marrying for sexual gratification, it will pass and there is suffering. The only marriage contract that works is what the original contract was—we enter into this contract in order to come to God, together. That's what a conscious marriage is about. In fact that is what everything you're doing is about. When you're ready you flip the figure and background, and what was figure becomes background and what was background becomes figure.

You look around and you find you have a whole set of existing relationships. Some of them are not based on sharing the journey to God, they have other

11

reasons and the reasons fall away and the relationships fall away because they were what you call friends and you outgrow your friends. They go on different paths than you do. That's reasonable. Other beings you are connected with, you can't outgrow: parents, children, relatives of one sort or another. You don't walk away. That is your given karma of the incarnation. You may grow at a different rate than they do and they become the fire of your purification. Because they will pull on you as you used to be and your job is to eat that. Until you get so even and clear that somebody can come up and say, "Hello, Dick." Or "Hello, Richard" or whatever. And I'm right there. "Yeah." Not, "I'm Ram Dass, now." You work with the karma that exists in your life space. Now, marriage is very peculiar in this situation because originally the marriage contract put your partner in the same relationship as a parent or child. It was not something you could walk away from, like a friendship. It was "until death do us part," and it became that kind of karma that you worked on. And even if your husband or your wife turned out to be the worst bitch or bastard in the world, that was your work! And if you really wanted to go to God, it didn't really matter. On the other hand you may have gotten into the present cultural position of seeing marriages as special friendships, or even not so special any more; people move in and out of them the same way they have friends, outgrow them and drop them. Now, in terms of the karmic situation if you have married unconsciously, you are faced with an unconscious predicament, and whether you stay with your partner or not is not as gross a karmic matter as if you have entered into the relationship consciously, and then break it off. That's a different matter. It would be the same thing with abortion. Unconscious people that don't understand get abortions. And the karma is reasonably light because it comes out of mechanicalness in them. They're not aware of what they're doing. They're functioning totally in terms of lust and greed and fear and so on. They're just lost. But once a being has looked up and is aware of his predicament, then one style of life isn't that different from another. It's all grist for the mill. They don't sit around killing that, and keeping that alive in order to make their life nice. They take it as it comes down the pike and work with it.

There is no form that in and of itself is closer to God. All forms are just forms . . . not better to stay single or to marry; not better to marry or to stay single. Each individual has his or her unique karmic predicament; each individual must therefore listen very carefully to hear her or his dharma or way or path. For one person it will be as a mother, or for another it will be to be Brahmacharya or celibate. For one it will be to be a householder, for another to be a saddhu, a wandering monk. Not better or worse. To live another's dharma, to try to be Buddha or to be Christ because Christ did it, doesn't get you there; it just makes you a mimicker. This game is much more subtle; you have to listen to hear what your trip through is, moment by moment, choice by choice. Is this one getting me closer or isn't it? And then you'll learn how truth gets you closer, how straightness gets you closer. You'll learn how simplicity of mind gets you closer. You'll learn how an open heart gets you closer. Certain acts—for example, like smoking pot—may show you the place, but over time they don't necessarily keep getting you closer. When you're finally really honest with yourself about it, you recognize that: that it showed you a possibility, but it doesn't allow you to become the possibility. In fact you get to recognize that happiness doesn't necessarily awaken you faster than sadness or unhappiness or pain or suffering . . . quite the reverse it turns out. Pain and suffering awaken you more, because the only reason you experience pain or suffering is because you are clinging to something or other.

When you have the compassion that comes from understanding how it is, you don't lay a trip on anybody else as to how they ought to be. You don't say to your parents, "Why don't you understand about the spirit, and why I'm a vegetarian?" You don't say to

13

your husband or wife, "Why do you still want to ball when all I want to do is read the *Gospel of Ramakrishna?*" A conscious being does all he can to create a space for going to God, but he does no violence to the existing karma to do it. So you work with your fire, but not patronizingly, because you're not superior; you're just different. When you understand about incarnation, and that surrounding you are beings at every level of incarnation; some of them very new beings who just started to take human forms and are very busy materially getting it together. While there are other beings around you that are very old who have been born again and again and again, and they've worked out an incredible amount of karma and are all ready to float into the akash, to float back into God.

Some of the beings around you every day are very ancient beings, and some are very new. But is it better or worse? It's just different. Is it better to be twenty years old than fifty? It's just different. So why do you judge someone because he's not as conscious as you are? Do you judge a pre-pubescent because he or she is not sexually aware? You understand. You have compassion. Compassion simply stated is leaving other people alone. You don't lay trips. You exist as a statement of your own level of evolution. You are available to any human being, to provide what they need, to the extent that they ask. But you begin to see that it is a fallacy to think that you can impose a trip on another person.

I used to meet people and I'd see how they could be and, because I'd done a lot of sadhana, I developed some minor siddhis or powers so I'd look into their eyes and touch them in a certain way and do a thing and they'd start to be who I wanted them to be. Then I'd say, "Wow, look at that," and they'd say, "Oh thank you, thank you." And they'd love me and want to follow me around. But the next day or week or month they'd come down because they were living out my mind, not theirs. They were living out my desire of how they ought to be, not being how in fact they needed to be in their own journey of evolution. The

14

best you can do is become an environment for every person you meet that allows them to open in the optimum way they can open. The way you "raise" a child is to create a space with your own love and consciousness to allow that child to become what it must become in this lifetime.

It's the same if you're a therapist or a marriage partner or a spiritual teacher; whatever your role in human relationships, the game is always the same. If you're a policeman on traffic duty, your job may be to give people traffic tickets. How you give that traffic ticket is a function of your evolution. You can give a person a traffic ticket in such a way that they'd end up enlightened. Because there is no form to this game at all. It's who's in the form that counts. It isn't how holy you look; it's how much you *are* the spirit of the living Christ, the compassion of the Buddha, the love of Krishna, the fierce discriminating wisdom of Tara or Kali. There isn't even one action or emotion that's holier than any other. People get into thinking one form of emotional action is more holy. It's like when Maharaj-ji says to me, "Ram Dass, give up anger." And I said, "Well, Maharaj-ji, can't I even use anger as a teaching device?" And he said angrily, "No!!!" There are many levels to this game; they sneak up on you.

Of course there are certain acts that conscious beings do not perform—not that they couldn't be performed consciously, but that it doesn't come into the flow. You will not be able for long to hide behind form. Many people say to me "Should I be a vegetarian or shouldn't I?" "Should I have sex or shouldn't I?" "Should I meditate forty minutes or shouldn't I?" People that meditate exactly the right number of minutes, eat exactly the right food, do all the things perfectly, can also be caught in the chain of gold, in the chain of righteousness and ritual. That is not liberation. But eventually you do perform the spiritual practices, not out of obligation, not out of guilt, but because you've got to do it. Because it's demanded of you by you. You end up going through hell in medita-

15

tion to quiet your mind, not because somebody says, "You ought to quiet your mind," but because your agitated mind is driving you up the wall and it's keeping you from getting on with it. You'll learn how to pray, and read holy books, and practice devotional acts and chants, opening your heart and asking Christ to fill you with love, not because you're good, but because with a closed heart you know you cannot come into the flow of the universe.

Now there are very delicate issues about passivity and activity and will and choice and so on. And you must listen very deeply within yourself. For you are continually making choices, and the choices boil down to going either in the direction of the harmony and flow and the will of God, the flow of the universe; or going against it. And you listen and feel that with the deepest kind of honesty you've got. This trip is based on just two very simple concepts. Total honesty with yourself, total honesty. If you make a mistake, admit it and get on with it. Don't cover your errors. The whole spiritual journey is a continuous falling on your face. And you get up and brush yourself off, and you get on with it. If you were perfect you wouldn't even go on a journey. Don't be afraid of making errors. You may choose the wrong teacher, you may get into a method that's no good. Many things can happen. Make an error, correct it if you can, without hurting another being's spiritual opportunities. There is another rule for this game: you may never use one soul for another. If your journey to God is keeping another being from going to God, forget it. You're never going to get there. It's as simple as that.

Listen to yourself and be honest with yourself. Those are the rules of the game. Listen inward and be honest. Now when you listen inward, you may not even know what to listen to. There are dozens of voices saying, "Listen to me, I'm the one. I'm the one, get all you can." See. "I'm the one. Give it all up." See. It's the superego and the id and all these voices all vying to be center stage. And you keep listening for what the Quakers call the "still small voice within."

You listen deeper and quieter, deeper and quieter—
the more you enter a meditative space, the clearer
you'll hear your dharma, your flow, your way home,
your route back to the source.

The Evolutionary Cycle

Back in the sixties when we gathered we were confused as to whether we were psychotic or spiritual. We needed to gather in order to reassure ourselves that if we were psychotic at least there were a lot of us. We were freeing ourselves from a cultural model of a reality that had been considered absolute. And as we started to break free there was much melodrama: violence, anger, confusion, as well as bliss and delight. Some of the confusion came because we kept trying to make the outside different as a reflection of the fact that the inside was changing. Part of that was pure in the sense that the new inner being was manifesting a new outer being, and part of it was impure because our faith was still flickering and we needed new symbols in order to reassure ourselves that we were in fact different. Some may recall the period when men started to grow their hair long and the power of that symbol, along with communal statements and alternative economics. During the sixties we were confused between internal freedom and external freedom, between revolution and evolution, because we didn't have models in our heads that would allow us to appreciate the grandiosity of the change that we were undergoing. So we kept reducing its implications and seeing it as a social, psychological, or political change.

During the late sixties and early seventies there was a period of fanaticism in our spiritual involvement. We were importing models from the East at a great rate and trying very hard to convert ourselves but, consistent with our tradition of doing things from the outside in, although we were taking on a lot of the symbols and accoutrements and might have looked like Buddha from outside, from inside we were just somebody who was trying to look like Buddha. We were very confused about vows and commitments, the relationship to teachers, the whole concept of Guru, and what the journey was about. In the sixties the word God was still taboo, so we talked about "altered states of consciousness."

Implicit in all that we were doing was still an attachment to the fact that *we* could do it, that who we were or who we thought we were could change ourselves and become whatever it was that Buddha was or Christ was. That is, we were living in a culture in which man ruled nature, within obvious boundaries, and we were so addicted to the rational mind and its power that we assumed we could think our way out of any predicament, we could figure out a new way to be through our thoughts and through our doing. But the predicament is that enlightenment is not an achievement; enlightenment is a transformation of being. And the achiever goes as well as the achievement.

Most of us didn't bargain for the implications of the trip we found ourselves on. We started to understand that it might have something to do with what had been talked about as "God" or a "coming to God" or, if you would rather deal with the unmanifest, the state of Nirvana. And we didn't really want them, we wanted to want them. That's a different level of the game. For most of us it has been quite enough to want to want God or to want to want enlightenment. That keeps us cool, safe, secure, with a feeling that we're moving in the right direction. It gets a little scary when you start to disappear into the Void. In *Be Here Now* we referred to it as "the crisp trip."

Now the strength in us lies in our honesty with ourselves about our predicament. We have tasted of something, we are drawn to it as a moth towards a flame, we recognize our own fear, there is less melodrama and dramatic histrionics and we are patiently and insistently doing the purification of being necessary for this transformation to occur. Realizing that we can't grab it, we tried that, nor can we ignore it, we all tried that. You try to grab it and up you go and down you come; another high to add to your collection of moldering butterflies. You try to push it away and go back to not remembering that there is something else, and you can't do it. As you're in the middle of your intense sensual enjoyment which you would like to get lost back into, there is always the voice which says, "You are now in the middle of your intense sensual enjoyment." You can't get in, you can't get out. And here we are.

The melodrama is passing away. We recognize now that we are bringing our external world genuinely and honestly into harmony with our inner perceptions, and we don't need to try so hard to create an external space to prove anything. We're learning not to over-kill with our intellect, to think our way into holiness, because it just ends up another prison, and we get caught pretending we're something we're not.

We are developing a deeper philosophical understanding of the predicament we are in as mutants, as evolving beings. We're listening inside to see what it is that is keeping us from that place or space or realization or connection that we have touched, tasted, felt, or somehow know about and we are starting to find the methods to get on with the work. We have begun to understand that, though we gather as a group and listen to one another, each of us is in a unique predicament and that you must listen to your own heart to hear what you need, you can't imitate anybody else's trip.

To characterize these individual differences in terms of evolution, let me share with you a model that is just a model. Imagine an evolutionary clock. At twelve o'clock there is perfect harmony, "the Tao" as the Chinese say, "the Way" Christ talks about. The perfect balance, the interrelationship of all things with nothing separate, each in its proper place. The tree is the perfect tree, the river is the perfect river, the human being is the perfect human being. All in its perfection.

At one minute after twelve, something is separate. At twelve it was the Garden of Eden: perfect harmony and balance. Then a bite of the apple and suddenly they're wearing fig leaves and God is asking, "Who told you you were naked?" Where did shame come from? It came from self-consciousness. And where did self-consciousness come from? It came from identifying with your thinking mind and thus experiencing yourself as separate from that which you think about. At 12:01 duality has been created: subject/object, thinker and that which is thought. Separateness.

From 12:01 to 6:00 there is a continuous attempt to solidify, protect, and increase the power of your position as a separate entity, to create security, gratification, power over the world around you, to recreate the feeling of well-being that existed when you weren't, but now you are. Who I am talking about is you; you understand what I'm saying?

Let's just imagine that 12:00 is a sort of total perfection; although it's obviously unlabelable, we'll call it "God" but since it really is unlabelable maybe we better just call it "G–d" so we won't get confused. Now G–d has within its perfection the freedom for any entity, such as a human entity, to pit its will against the total will, or G–d's will. So at first there were beings pitting themselves against the system, against the harmony, then everybody was "us," and "them" were the forces of nature, the storms and so on. But between 12:01 and 6:00 a bizarre thing happened. Slowly "them" started to become others of us. Our tribe was "us" and other tribes were "them." Then within tribe there was the family and pretty soon it was "our family" and everybody else was "them." And then, within the family, Uncle Dave screwed us on that business deal so he was sort of "them." You couldn't really even trust the greater family that much, "us" had to be your immediate family. That's around 4:30 or 5:00. Then there was a generation gap you can't trust elders or youth, so maybe it's just me and my wife or me and my husband. And then there's a sex difference so I can't fully consider my spouse as "us", then I'm "us" and everything else in the universe is "them."

"I'm very strong, I've got my protection, I know where I am, see?" You think it would end there. But that's about 5:45. In the final fifteen minutes is what now is called the total alienation of an individual. From whom? From himself. So finally you're looking at yourself from outside and you don't even trust yourself so you're "them" too.

And what was the greatest power we had to work with in this journey from twelve to six, what was the greatest siddhi, power, that was available to us as long as we were attached to our senses and our thinking mind? It was our intellect. Look at what our intellect has done. Look at this illusion. Look at the awesome impact of technology. They are all extensions of the human mind. At the moment I'm living in Manhattan where, except for Central Park, there isn't anything you see that hasn't been run through a human mind. It's living inside human intellect actually. And the power of the human intellect is based on discrimination, individual differences; if you can tell the difference between this and that and you can do it better than anybody else, you get paid more. And this intellect, which now decided that it could do anything, started to create models of what it had to do in order to get into that space it remembered somewhere inside of itself as that perfect feeling of at-AUM-ness, perfect well-being. It developed a number of strategies. The most obvious one in our culture is "more is better."

Most of us have been on that trip, haven't we? On the supersensual astral planes. "Have you heard that new record by the blups? Yeah, but have you heard it when you're in the bathtub—with somebody else? Have you heard it when you're in the bathtub with somebody else by candlelight? On a good stereo set? There's an incredible wine; put it at the side of the tub: musk oil in the bath, the incense, the candlelight, the wine, the other being and the bath water is just right and on the stereo ... Man ... Oh ..." More is better. The obvious predicament that the intellect has a difficult time with is the sneaking realization that more is never enough. Or, more is maybe enough for a moment but it doesn't last.

If you watch the patterns of your desire systems and mind, the end of your day goes something like: "I think I'll take a nap. Gee, I'd like a cup of tea. How about a cigarette with that? I'm gonna listen to that music. What are we gonna have for dinner? What do you want for dessert, ice cream? I'm gonna have some coffee. What's on television? No, let's go bowling. Bicycle? Great. Ice cream soda? Let's go home. Okay. Want to go to bed? Okay. Ah, that's great. Got a cigarette?" On and on and in the middle of your main course, you're already thinking about what you'll have for dessert. The way you deal with this game is by constantly keeping the things going by fast, like a sleight-of-hand trick. Knowing that none of them will last, you figure that enough of them with small enough spaces in between will keep the rush going. Rush after rush after rush. But it's like building a house on sand—you can't stop because it gets a little frightening if you stop. If those spaces in between get too big, there is depression, confusion, disorientation, anger, loneliness, self-pity, unworthiness. Such stuff! Yech! So keep it coming, Ma. More and more and more.

But it turns out that Christ was right when he said, "Lay not up your treasures where moth and rust doth corrupt and thieves break in." Buddha was right when he said, "The cause of suffering is craving," craving after things that are not permanent, and nothing is permanent. So if you cling to anything which is in form you're going to suffer. That was Buddha's point. What is blue chip, that you can invest in, that you can stop feeling frightened about? Your body? Your body is decaying at this very minute. Even the youngest person here is decaying. Fifty or sixty years from now, you know where your body will be, what it will look like? And your intellect? All the knowledge that you've collected? Did you ever see a skull and consider what's been eaten away and who ate it? And do you know what that emptiness is? That's everything you think you know. No wonder you're frightened. If you're thinking you're your thinking mind or that you're your body it's panic.

From 12:00 to 6:00 is the increasing hope that

you can get it all together, get it to feel just right. But there's a scariness because you're trying to do it in a dimension that exists in time where everything changes, you're going to lose everything. At the very least you're going to die. As a philosophical materialist—not a materialist in a gold Cadillac, but one who is attached to the senses and intellect, and what you can think about—you are afraid because when you're dead, you're dead. As you get close to dying, you start to get very frightened and you start to push pretty hard. You say, "Doctor, you've got new pills, use them, do anything, save me, freeze-dry me, do anything, I don't want to die," and grab and hold the bedsheets and pay more and more and get more and more hysterical and get into intensive care units and keep alive even if they have to transplant everything. But no matter how hard you try, suddenly you're dead.

And then a voice says to you, "Hello." If you're a philosophical materialist, this leads you to say, "I guess I didn't die." To which the voice replies, "Oh yes, you did." Just as an example, at one point Buddha with his clear vision looked back and saw his last ninety-nine thousand incarnations. And that was only a trivial amount of them. Birth-death-rebirth-redeath, on and on.

Now, in the early period, say between 12:00 and 3:00, every time you die you're so caught in your own attachments to your senses and your mind, you're so deep in the illusion, that when somebody says you're dead, you deny it and stay in total confusion until you get sent into the next round. Which is all, as you will see, perfectly designed. Later, as you get on with this round of births and deaths, you realize your predicament. You are under the veil of illusion of the birth—you don't want to die, then you're dead and you say, "Far out, there goes another one." At this point you look around and you see all your old mothers and fathers and friends. "Oh, my God you were my wife this time, last time you were my brother."

When you're more conscious you share in the understanding of where you are situated on the clock, in the round of births and deaths. You begin to see

24

exactly what the next birth has to do from a karmic point of view, what it has to work out. And when you design the next birth, you say, "Well, I think I should be born into the lower-middle class in New York City, and then around ten I think I should get raped, that would be useful for that particular samskara, that deeply imbedded mental impression, that I've been working out from four thousand births back. Let's see. I'll have my first child when I'm eighteen," etc., etc. You design it all the way through—up until how you'll die. You've run it all through the computer, the right parents come together, the right combinations have come up, comes the moment of birth. And there you go. You dive back down in.

Some beings enter into this trip at the moment of conception, others at the moment of birth. You can tell those babies that entered at the moment of birth—the baby comes into the world and has that kind of stoned out look, like what the hell am I doing here!? Like an old lama who has been born, say, in the Bronx, and he would like to bless everybody but he can't get it to work. The ones that entered in at the moment of conception are busy being babies already. "Waaaaaa, give me." Those that come in stoned, because the veil hasn't shut down yet, are most of the time around parents who are busy inside the veil saying, "You're a baby, you're a baby. Goo-goo, look at the little baby." Pretty soon you buy it, and there you are again. On and on and on and on. Until something interesting happens at 6:00 o'clock, or one minute past six. Up until then, in every birth you've gone back into the illusion that you are this body, you are this thinking mind, you are these senses. Everything you could get is what you can sense and think about, you're grabbing, looking out and down. Grabbing and grabbing and grabbing and then suddenly at one birth there is a moment when the veil parts albeit for just a second and you stick your nose through and you say, "Wow, it isn't how I thought it was at all." Maybe the veil parted for a millisecond, but that was all it took if you were ready. The veil is parting all the time for everybody, but most of the time your karma is so heavy, and you're so used to the veil, that you're not ready, so that the minute you do see through you immediately deny it or push it away as hard as you can. Recently I read in the *New York Times* magazine an article on "Mysticism in America" which said that two-fifths of the population of the United States has had a genuine mystic transcendent experience, which means they saw through the veil. In sampling that two-fifths of the population, eighty-five percent said, "It was the greatest experience of my life but I never want to have it again." Of course not, because look at how it screws up the apple cart. If you've built a whole universe around being somebody and suddenly you see that that isn't who you are at all, what then?

But what is the condition necessary so that the moment comes when you see through the veil in a way that changes everything from then on so that from 6:01 to 12:00 your whole journey changes its meaning? The condition necessary for that to happen is that despair, the realization that everything you think you can do to create perfection isn't going to be enough. That who you are and who you think you are is where the problem lies. It leads to a deep despair that is absolutely the necessary condition for you to look up at that moment. Once you have seen and know you have seen, you can never totally go back to sleep again. Even though you may forget for moments—and you will go through many, many more births from 6:01 to 12:00—you can never fully forget. You are starting to be drawn back to twelve o'clock.

I'm talking about a clock of births and deaths which is all in time, which is all an illusion or relatively real, but we're just working with this illusion for a moment. The nature of the beings in tune with these words are by definition after six o'clock. Otherwise there'd be no reason for you to have read this far. Maybe you're a 4:13, but why would you put up with this long rap then when you could be out getting more which is better. But you know something and you're trapped in what you know and look at what it's led you to. And it gets worse; that's what's so far out about it: Once you start at one minute past six the return journey to 12:00, you're trying to grab at experiences that are going to get you back. You're going to collect new experiences that are called "getting high." You come down from something and you treat that down as the time between the last time you got high and the next time when once again you'll get on with your journey to God or back to 12:00 or whatever you want to call it. As the experiential clock ticks on you keep developing in your understanding of how it's all working, and you begin to recognize a peculiar phenomenon that, as C. S. Lewis points out, "You don't see the center of the universe because it's all center." That you, in fact, are the center of a universe which has been designed perfectly in order to awaken you out of the illusion and that every experience you have is equally valid as grist for the mill of awakening. Your whole incarnation is the teaching.

Next you begin to realize that although they are all equal in teaching quality, some of your experiences seem to shake you more than others, that the model which you are stuck in, sometimes so subtly you don't even know it, is shaken by pain and suffering and all the negative qualities. At that point you recognize the bizarre phenomenon that suffering is grace. Now that's heavy. Because up until that time you've been trying to optimize pleasure and minimize pain. When you realize that in its fullest dimension, you may still live to optimize pleasure and minimize pain but whatever comes down the pike is all right. "Boy, am I depressed. Far out. Now, there's depression." Until finally "There's pleasure." "There's pain." "I just made a thousand dollars. Wow." Or, "Oh, I just got robbed." And the "Wow" and the "Oh" and the "Ah" and the "Uhh," all of these alternatives are just more stuff, beautiful, delightful stuff. This incarnation is the absolutely optimal one which you must be in now in order to do what needs to be done, or have done through you what needs to be done, in order to bring you home, bring you Aum, or out or in. It's happening whether you know it or not. But as you know it, it changes it, that's part of it, that's all karma too.

Along around 10:00 or 11:00 you're going into

27

other planes of reality in your meditation or however, and they are equally as real as the plane that you started this incarnation in. You don't quite understand where you're at, sometimes you get confused, it's very uneven and complicated work. But if you're really aiming at perfected truth, you move at a rate at which you can keep it all perfectly together. You work for the perfect balance of the different planes. At one minute past six you started to look up and got so fascinated with what you started to see you couldn't take your eyes off it, and forgot to look down and you fell on your face. You started to study the "absolute truths of raw God," and got so fascinated with the impersonal perfection of the universe beyond all polarities, you were so involved in the icy cold impersonality of it all that you kept stepping on things and you looked and said, "Well so what, it's all perfect." But you learn the simple rule of the game is that as long as you push away one plane to grab another, you're still off balance. Ultimately, you understand that the truth must be balanced with the caring, with the honoring of this incarnation. That's when you start to develop the capacity to look up and to look down at the same moment. To look in and to look out.

When you look at pure truth you can see the grace that suffering is. From your point of view, when you're suffering, "Fine, I'm suffering, that's interesting," at the same moment if you are looking down and honoring your incarnation, you're working to alleviate suffering. Let me give you an example. Somebody says, "I want to study yoga with you. I want to fast." And you say, "All right, fast for nine days." At the end of the seventh day they say, "I've fasted for seven days." And you say, "Wonderful, wonderful. You've got two more days to go." Then you walk outside the building and somebody comes up to you and says, "Hey man, you got a quarter, I haven't eaten in seven days." You don't say, "Wonderful, wonderful, you have two more days to go." It's not an appropriate response, because for that being, suffering isn't grace, suffering is a drag.

When that discipline is developed to allow you to

look up and look down simultaneously, you have the absolute clarity of the pure white snow on the Himalayan peaks, the exquisite clarity, raw truth, the impersonal perfection which includes everything— Vietnam, Cambodia, Bangladesh, Biafra, persecution in our cities, inequality, violence, as well as all the bliss and love and compassion and kindness—the entire mosaic. In the icy peaks of the Himalayas you see the perfection of it all in the evolutionary journey of beings. And at the same moment the caring part of you is like the bleeding heart of Jesus and you look down and see the blood on the snow. You keep both of those in mind at every moment, so you can help beings that are suffering in the way they need to be helped.

If you are really going to help them get out of the illusion, you yourself must not get lost in the illusion, you must continue to keep your eye fixed on absolutely clear truth. You love without clinging, you help without being identified as the helper, you protest without getting lost in your protest, you care for your child remembering that, behind it all, here we are. The truth, the caring. You honor your body, you honor your society, you honor your whole game, you change it in the way it needs to be changed. You listen to hear what your particular karmic predicament is in this round and you find your dharma, the way to live this life in perfect harmony with the forces inside and outside of you in order to bring you home.

If you get greedy and try to push or pull, you're going to fall on your face. If you go up to sit in a cave, you'll become so holy light will be pouring out of your head, everybody will be falling at your feet, and you'll have great powers. But you try coming into New York City and you will see that there are little seeds inside of you, as Ramakrishna talks about, that never quite got cooked. It's an interesting point of view when you say, "Hey, I can't stand to live in the city, I've got to live out in the country." All you're saying is, "I can't stand those things in myself that the city fans." Believe me, if there's nothing that you want, the city is the same as the Himalayan peak. All the city is showing you is stuff in yourself that you wish you didn't have.

As you get further along in this journey, the pull of twelve o'clock gets so fierce, you want to get done so bad you can taste it, and at that point you say, "Give me the fire. I want a hot fire. Make it hotter, hotter. Come on, give it to me." Then when somebody gets you furious you know that the only reason you got angry is because you have a secret stashed model of how you think it ought to be, that you're holding on to. You realize that the person that got you angry is a teaching and in your mind you thank him. You get so eager to root out the stuff in you that's keeping you from getting on, from awakening, that you start to look for situations to force you to do it.

A couple of years ago I spent nine days in a seshin, a Zen Buddhist retreat. It was without doubt the most miserable, horrible, cruel, sadistic experience . . . I got sick, I was paranoid. They sucked me in, they seduced me on my ego, making me feel like I could do it. Then, I got there and they didn't even give me a reward of saying, "Ram Dass, welcome." A guy met me with a clipboard and said, "Dass, Ram; you'll be in the upper bunk in Cabin Three. Here is your robe. Report to the zendo in five minutes." There was a guy with a stick and unless you sat in perfect form, which was really uncomfortable, you got beaten. And I was paying money for this! If you tilted, this really fierce character would come up and he'd bow to you, then you'd bow to him, and then you'd lean over and he'd beat you on this shoulder and then you'd lean over and he'd beat you on that shoulder, and you'd thank him and he'd thank you. Five times a day, you'd go in to see the Roshi, a tough Japanese fellow, bald headed. He had a bell and a stick and he'd ask you ridiculous questions like, "How do you know your Buddha-nature from the sound of clapping hands?" And you'd answer something or other which you'd been thinking about the whole time you were sitting there, knowing you had four more to go yet today. And he'd say, "Oh, Doctor, you're not doing it right at all. Maybe we should give you your money back, you leave. I had great hopes for you. You're very important, people know you, you're very famous, but you don't seem to understand this, I think you'd better forget it."

Then he rings his bell and you leave and you're crushed. Not only are you crushed but you have to run back to the place to sit up so as not to get beaten. This goes on from two in the morning until ten at night. There's no edge. I spent five days plotting how to be called away on an emergency, some face-saving device. I even tried to hide in the bathroom, but they checked the bathroom. There was simply no place to hide.

Finally, by the fifth day, I didn't give a damn about the Roshi or the whole scene. I went in kind of slouching, thinking "The hell with it, let them throw me out," and he said, "Doctor, how do you know your Buddha-nature through clapping hands?" And I said, "Good morning, Roshi." He said, "AH!!" He was delighted, and smiled and then, lest it go to my head, he said, "Now you are becoming a beginning student of Zen!"

Well, it was interesting because just before that, just as I was walking up the path and saying "Screw it," fire started to pour out of all the bushes and the whole sky became radiant, and I went into this other state. It was like I had been released from this incredible sickness and tension and I went in and I was having a satori experience. And he kept asking me koan after koan and the answers kept coming right out. I was right in the moment and there were no models in my mind. And we just went higher and higher together and we were both just spinning out.

From then on the rest of the nine days was ecstasy. The sittings were beautiful and I was just floating. Suddenly, the perfection of the emptiness of the forms and the impersonality, became my freedom. Had I come to the seshin and had they said, "Oh, Ram Dass, we're so happy," and had I known who everybody

31

was, it would have occupied my mind in a way that I was freed from by the total impersonality of the whole scene.

It's just like any meditation when it's not all bliss and light and you're uncomfortable and it's hot and you're bored and your butt hurts and all that. It's all the same as that seshin. But you do it because there's something you want bad enough to go through it, to struggle against the forces in you that just want "more." That's what the *Bhagavad Gita* is about, the battle inside between those two forces. Right until the very end it's hell. It doesn't get any better, it gets worse because the fire gets hotter and hotter.

You see, once you decide that you really want to go for broke, for perfected truth, once you're being pulled that way in your gut and you finally say, "I don't want anything else, I just want to go," (which is usually a lie, but you're still saying it) that pull, that reaching, draws down upon you all kinds of forces which help that thing happen. That's called grace. There are many beings, both on this plane and on other planes, that are available to guide you and help you, but they don't come unless you want them. Your reaching elicits their help.

The teaching gets fiercer, the fire gets hotter, you start to do it to yourself because the pull of God is deeper and deeper. At that point just before twelve on our evolutionary clock the entire universe is within you and you experience all of the suffering that is connected with form on any plane of existence. You are one with it. By that time, you have worked out all your personal karma or clinging. Now you're aware of the collective nature of the karma. Right at that moment the pull to twelve o'clock is incredible. To go into twelve o'clock means that you merge back in, you as a consciously separate entity cease. Everything that went on from one minute past twelve until 11:59 was designed for this moment of choice. If you want to be God at this moment, you can merge back in, but it doesn't matter what choice you make. That may be scary because you wanted it to matter. Most of us are so caught in righteousness we're afraid of truth. Righteousness would say it matters at 11:59 but truth says it doesn't matter. At 11:59 you have the choice of going back into God in which case if you had left a body it would just sort of disintegrate because there was nobody in it, or you can stay back in form on this or another plane. Why would you do it? This is free will in the true sense of free will, not the illusion of free will that we have, for there is no individual karma in this. The only reason a totally free being would choose to stay within the illusion is in order to relieve the suffering of all beings. This is the time when the vow that's known as "the Bodhisattva Vow" is taken. This is the only moment it's real, up until then it's phony, it's your karma working out. The moment you choose to come back you have to push against that force that is drawing you in to merge. You are pushing against God. That is the sacrifice. The sacrifice that Christ made is not the crucifixion. The chance for a conscious being to leave his body is bliss. The sacrifice was leaving the Father in the first place and becoming the Son. Free beings, realized perfected beings, have that free choice. They are here only because of you, and I mean you, because otherwise you wouldn't meet them. Anybody they meet is by design part of what they're doing to relieve suffering. They are here only as instruments to bring through that non-clinging, non-attached truth, to create a mirror against which you can see where you're holding your secret stash of stuff that is keeping you from being perfected also. These are the beings that bestow the grace. They are the Gods and Goddesses and Gurus on all planes. Everyone has one of these helpers specifically designed for your karma, but most never meet them in this lifetime because they never reach out.

Every night Buddha would look over all the realms, the Buddha-fields, to see who was ready, who looks up, who is reaching, who says, "I want to get out," who says "Know me, let me out, I'm ready, let's go." Not wanting to want, not phony wanting, but wanting. If you don't reach out nothing happens. For whom is the despair deep enough?

This game is designed so that within the illusion

where there is free will, you've got to reach for it. And you only reach for it when your karma allows you to reach for it. See the predicament? The only real free will there was in the whole clock was at twelve o'clock to one minute past twelve—the free will to go against the system—and at 11:59 to go back into the system. Otherwise all of it was determined by law.

Keep in mind the entire clock is in the realm of illusion, or relative reality. At twelve o'clock you never were, nothing has happened, nobody is. To answer the question of why did it all begin, one of the answers is it never did. It's just a play of mind, just a play of mind. People who have come this far in this transmission are everywhere between one minute past six and 11:59 and because of the nature of your attachments you can only see what you can see. You might be sitting next to an 11:59er and you wouldn't know it because he doesn't have a sign on him and the ones that do have signs on them usually aren't because they wrote them themselves. It might turn out that your Aunt Thelma was Buddha. She was cooking chicken soup and you went to India and Tibet for forty years looking for somebody that looked like Buddha. You got totally despairing and in the despair you gave up all your hope and all your models. You came home and you walked in and there she was. You look and you fall on your face before this brilliant light and she says, "Have some soup." The pure Buddha, the mind that is clear of attachment, exists anywhere in perfect harmony with all the forces around it.

And to complete this clock image I might add that for some of you it has become time to awaken and for others it's later than you think.

Levels of Reality

It's useful to understand the various levels of reality, to examine the perceptual fields that different beings have, to see what different realities look like. Imagine that you have a little dial right next to your eyes and that you can change the channels of your realities. These channels of course are not to be confused with the chakras. Set it on the first channel and you look around the room and you see men and women. You see that some are tall and some are short and some are light and some are dark, some are beautiful and some are not beautiful, some are blonde and some are brunette, some are fat and some are thin, some turn you on and some don't. That's the physical reality. Somebody says, "Who was in the room?" You say, "Well, there were about an even number of men and women. And they were mainly young between the ages of . . . ." If you were a social scientist you might say, "There were so many endomorphs, so many ectomorphs, and so many mesomorphs." If you're in social action, you might say, "There was a minority of blacks, there were so many Protestants, there were so many etc." If you were primarily sexually oriented within that domain, you would see everybody in one of three categories, either potentially makeable, a competitor with you for someone who is potentially makeable or irrelevant. And that's a very dominant theme in the *Playboy, Penthouse, Oui* clientele which involves a large percentage of this society. That's what's real for them, the rest is all trips. When the dial is set on the

35

first channel, when you look at the world, you see the physical, material environment.

If you're in the clothing business you walk down the street and you see what everybody's wearing. That's the reality for you. And if someone asked, "Who was it that just passed?" you'd say, "I passed a gown from Bergdorf's," or "shoes from Saks and a hat from Filene's that was on sale last week." Or if you're pre-occupied with your body when you walk down the street, you know what you see? Everybody else's body. People who are busy being short are preoccupied with how tall everyone is. People who don't like their noses notice everybody else's nose.

Flip it one flick, the second channel, and we're in the psychological domain. If you were a very technical person you would now look at everybody in terms of the Minnesota Multi-phasic Personality Inventory or the Rorschach ink blot test. We are now looking at happy, sad, achievers, anxiety neurotics, manic-depressives, enthusiasts, spiritual seekers. Eager, depressed, hearty, happy, sad, lucky—a whole lot of psychological attributes. For many of you that is the reality in which you live. And who you are is your personality. You spend time analyzing and therapizing it, patting it, damning it, feeding its guilt, its shame, its unworthiness. It happens that the psychologies come in bodies, but you don't even notice the bodies, you're too busy with the psychologies and personalities. When you meet people you say, "Our personalities do well together." It's the only reality. They don't notice bodies. They don't look up, they don't look down, they're all psychology. That's the personality level.

Then there's another channel. Flick. Now the world is twelve categories and their various permutations. There's a Leo, there's an Aries. "I know you're a Sagittarius, I can tell by the way you walk." You've now done an astral fix, a new game of individual differences. You now know people's subtle body which is inside their physical body and you know about something that lies behind their personality, a planetary reality. It's another game of individual differences. But the third channel allows you to re-perceive the first and second channels. You're now using one set of individual differences to free you from another.

What do you say, once more? Flick, fourth channel. Now when you look into another person's eyes what you see is another person looking back at you. "Are you in there? I'm in here. Far out. How did you get into that one?" Now we see another being who is just like you, another entity trapped inside the illusion of all these packages of individual differences, body, personality, astrology. And the eyes, the windows of the soul, meet and you say, "What's it like being in there?"

You'll notice that when you meet other people, you're meeting them at all these different levels. If, for instance, you're a girl and you're on the personality level of reality but you happen to have a very beautiful body from a cultural point of view and everybody you meet sees only your body, you might say, "Why doesn't anybody want me for my personality?" That's because you're such a strong stimulus on the first channel nobody can get to channel two. Or you're sitting on the fourth channel, which is just a soul inside all this stuff, but most people are busy responding to you as a personality, a body, an astrology. Though every now and then you'll walk down the street and you look into somebody's eyes and there's somebody else who's looking back at you. They're not coming on, they're not trying to seduce you, change you, buy you, collect you, proselytize you, reject you, judge you or anything. They're just there. "You're here, I'm here. Far out place to meet isn't it." It's just beings meeting inside these packages of individual differences. And no longer are these individual differences the reality that is so solid. They're just like shirts and jackets and sweaters. "That's a pretty personality you're wearing. Where did you get that one?" "I bought it in Gestalt Therapy. It's primal scream." But it's still separate. You're still separate from another.

Once more. One more Flick. Now what do you see when you look at another person? It's as if you had two mirrors facing each other with nothing in between.

37

It's itself, looking at itself looking at itself. In that reality there is only one of us here in drag. One is appearing to be the many in order to play out this game. We are all the Ancient One. We are the One. And the One becomes the many for the play, for the sport, for the dance and you can get lost in the many in realities one, two, three or four; but in channel five, there is only one of us, not intellectual or metaphorical, in that reality, we *are* one. Every reality up to this point is an equally valid, relative, symbolic reality. They're all real, but they're all just relatively real. One of them is no more real than any other one. The way in which we're the One is no more real than the way in which we're the many.

Flick to the next channel and what happens is that everything out there disappears and you disappear and there's nobody looking at anything. The whole television disappears, it all goes back into the Void from whence it came, it returns to the formless—to that which lies behind the One. It is what God is, not the concept of God, but God Itself. It is in Buddhism, Nirvana; it is in Hinduism, Brahma; it is in Taoism, The Eternal Tao. It is the Aum, the unmanifest universe. It is why the Hebrews write God as G–d, because it isn't speakable, it is the unspeakable source of it all. When you flip to the sixth channel, the whole subject-object universe dissolves. And all just is, but without form. Because in order to know form you have to be separate from it. To enter into the formless that lies beyond the form, out of which the form comes, and back into which it returns, is to touch upon the reality out of which all relative realities arise. As you enter into the realm of channel six, the last moment of recognition of any kind of self-consciousness is the realization that all of channels one through five were creations of mind. A liberated being is a being who is free to be in, but not attached to, any of the realities. It is a being who can enter into the ocean of channel six and yet return into form. It's a being who has all of the channels available at once, though he may attend to them somewhat sequentially.

Every one of those flips is a reality. But when

you're totally in one of those, it is your absolute reality. You're somebody sitting here reading this book, that's a reality. But is that any more or less real than the reality that there is only one of us reading itself? It's just a different level of reality, a different flick of the dial.

Part of the process of awakening that we're going through is the recognition that the realities which we thought were absolute are only relative. Since as we flip the dial we go towards more and more energy, and more and more fineness of vibration, when we meet a new reality we experience an intensity that makes us think that the new reality is more real than the old reality. We are living simultaneously on a number of levels. You are existing and functioning simultaneously on many levels, living out the karma of your separateness on plane after plane after plane. But you as a self-conscious entity are identified at any one moment most likely with only one of those planes. Thus to define yourself as being within any one plane of this baklava is to impose a limiting condition, and you are then less than free. And even considering channels one through five as real is just an imposition of the intellect, describing structures. The intellect goes up only a few levels, and then it becomes a limiting system. When you go back about three or four planes, there are no time or space dimensions. Past, present, future, here and there, are all here. The Buddha looked back over his last ninety-nine thousand incarnations, just like that, and saw them all clearly and simultaneously, because he did not have to be limited by his linear mind. So part of the key to acknowledging your other identities is the process of allowing other kinds of knowing to be real for you, other than the ways you know through your five senses and your thinking mind. Sometimes we call it using the intuitive mind. Heinlein called it grokking.

Now the fact that you were born into this plane that exists in this place at this moment means you were born into channels one and two though you also exist on channels three, four, five and six. Most of the people around you that you have grown up with take

39

channels one and two as absolute reality. Thus when you go into channels three, four, five and six, they say, "Come back to reality. Get your feet back on the ground. Be realistic. You aren't being realistic."

People who acknowledge channels three, four, five and six are often treated as psychotic. "You have flipped out of conventional reality." For someone who sees through the game of relative realities, you see that the clinging to any reality at all as "the reality" is really the definition of insanity. I once visited a brother in a mental hospital. I sat in a room with him and his psychiatrist. He thought he was Christ and the psychiatrist thought he was a psychiatrist, and each of them was convinced that the other one was insane.

Most of the experiences that you have sought in your life have been an attempt to enter the universal oneness of channel five—total orgasm: the moment in which there is no longer somebody having sex with somebody, when it's merely the universe happening, that moment of perfect flow in which all separateness has disappeared, is the moment when you are home again, when you know where you belong, when you have returned into the One, when all of the tension that is created by the separateness has for a moment been dissipated. And most of you know that at the moment of orgasm all of your neurosis is irrelevant, not the moment before or the moment after, but for that exact moment. For somebody who is capable of living in channel five, that moment of orgasm, of total merging with the One, is a reality all the time.

In order to be open to this merging, many of you that have smoked dope or taken acid, or had other vehicles for overriding your programs, know that you can set aside your program for a moment and enter into the higher channels, but after a while, you come down and as a result get very frustrated. And what brings you down is your attachment to the models or molds or programs about who you think you are, and how you think the world is—these habits of mind.

Many of us are getting to the point in our spiritual journey where we are no longer trying to get high, for we know how to do that, we are trying to *be*.

And being includes everything. We now recognize that if there is anything at all that can bring us down—anything—our house is built upon sand, and there is fear. And where there is fear, you aren't free. Thus you become motivated to confront the places in yourself that bring you down; not only to confront them, but to create situations in which to bring them forth. That's quite a flip-around from a mentality which says, "I just want to get high." That's the mentality that says, "I want to get done; I want to be liberated in this very birth. I've seen how it could be; I'm tired of just seeing previews of coming attractions; I want to become the main feature."

That mentality is quite different from the mentality we had ten years ago. For now when there is depression, instead of running and hiding from the depression and trying to grab the next high, you turn around and you look at the depression as though you were looking the devil in the eye, and you say to the depression, "Come on depression, do your trip, because you're just a depression and here I am." Because you are just a little bit connected to channel four which lies beyond channel two, and the psychological channel is where depressions are.

To a being who is living on the fourth channel, what do you suppose sexual anxieties are all about? "Was I good? Was it enough? Did I satisfy her or him? Will it happen too soon? Can I get it up? Should I fake it? Am I frigid? Will it be real?" Those worries are on the first and second channels. They're not, not real; but how different it would feel if your identity was rooted in channel four. If you were able to say, "I am a soul that has taken incarnation in a body that has this particular sexual habit." You even arrive at the space where you are able to look back at your entire life of neurosis and suffering and say, "Look at how perfect that has been in bringing me to this moment."

So we have the paradox that on channels one and two and three there is an incredible melodrama going on, and you are an actor within it. There is a great deal of suffering, and when you're locked into channel one and you're hungry, and there's no food; that's real suffering, it's not fake suffering. But from channels four and five you can look at one, two and three and all the individual differences and all of the melodrama and say, "Look at the perfection of the dance. Look at the perfection of the flow of the Divine Law—including the will of man which can go against the Will of God. Look at the perfection of it all."

But it's not freedom for the being attached to channels four or five who, when you are hurting, says, "It's not real, don't worry, you're the Buddha, it's all an illusion." That's no more liberated than the person who's caught in one, two and three and says, "There's only pain and suffering everywhere and it's hell, life is terrible and ugly." The free being lives within all those realities simultaneously. He understands paradoxically that when there is suffering, you must do everything you can to alleviate it. While at the same time, the suffering is perfection itself and your doing everything you can to alleviate it too is part of the perfection. Because ultimately you understand when you are free to move without attachment from level to level of reality, that the only reason you stay in form is to alleviate suffering or bring others to the light, to consciousness, to liberation, to God. What a paradox! Is it perfect? Sure. Is it hell? Sure. Both at the same moment.

And all this too is at only one level of exploration.

The Mellow Drama

There will come a time in the not too distant future when we will be able to gather and sit in silence, not in expectation, but in fulfillment. For we will recognize who we are, that we are beings of the spirit, and we will seek that food which feeds our soul. Our intellects will be available, but at rest, and our hearts will be filled with the love of Christ, that flowing, conscious love. We will be done romanticizing our own journey, examining ourselves self-consciously to see how we're doing. We won't have to compare, assess whether we're getting enough, for we will trust our hearts.

For many of you, that time is now. For others, the faith is still too flickery. Recently in lectures I've spoken a lot about phony holy—that is, appearing holier than you are. Here I would speak more about phony unholy. Many of you are higher than you are copping to, you create molds or models of your reality that keep you from recognizing yourself. For many of you your spiritual model was "the good life." "Good" meaning a life consciously lived, simple, righteous, not ripping off people or the environment, socially conscious, keeping your scene straight, and being a person of peace. Not lost in your frustration, in violence, in anger, in lust. That's pretty good. To have found that level in this society puts you already in the tiny part of one per cent. But even when you get all of that together, there's something in you that's yearning. Because no matter how subtly beautiful your model is, it still defines you within worldly concepts. But in truth though you live in this world you are not solely of this world. You have learned to live in this world, but you have not yet fully recognized where you come from, or who you are. For to explore beyond the world is a bit like stepping off into the void; it's like diving off a diving board when you're afraid of diving.

But if you're asking for freedom, if terms like liberation, or realization, or enlightenment, or living in God have any meaning to you, then leading the good life is just part of the way home. Many of you, in this very lifetime, have come from tremendous pre-occupation with your neurosis, with your achievement, with your career, with your melodrama, and you've arrived at a place where you can laugh. You can say, "Isn't it a mellow drama!" Not just someone else's, but your own. Some of you are not as busy being neurotic as the general population, because you've begun to take personality a little less seriously. It's there, just like your body is—you comb it, wash it, clean it up, dress it, move it here and there, love it, caress it, stimulate it, get rushes off of it. It's your traveling temple. Your personality too is just another shawl, a cloak. But there are a good number of you who realize that you are not just either body or personality.

We have been together for so many years now, going through so many trips together. And I've hung in, as many of you have. In most lectures, now, I have given up talking about almost all of my personal history, not because don Juan told me to, but just because it's become a kind of limiting condition. I'd rather now just talk about how it is, when you are free to play with God for the rest of your life.

But still some things need to be said. I'd like to just play out the "his-story" a little bit. In 1970, I had been back from India about three years, *Be Here Now* was just about to come out and I was pretty freaked by how much I was lost in the world. I went running back to my Guru in India. When I arrived there, he asked me, "What are you doing here?" And I said, "Well, I'm not pure enough to do whatever it is I am supposed to be doing—I don't even know what it is,

44

but I'm not pure enough to do it." He hit me on the head, pulled my beard and said, "You will be."

For a year and a month, I followed him around India, and every time I'd go to see him, he'd throw me out. He'd let others stay with him for months but I'd come and he'd say, "Jao! Go away! Go to Delhi." Go here, go there. I had many adventures and each time I'd come back and I'd say, "Maharaj-ji, you promised you'd make me pure enough." And he'd just laugh or say, "You will be."

Finally, I was being thrown out of India by the Indian government because of visa problems—Let me explain that when I got to India that time, Maharaj-ji said, "How long do you want to stay?" I said, "I don't know, I want to stay forever." Which wasn't true, but I thought I should say it—you can only take so much dysentery. And he said, "How about March?" This was February. I said, "You mean a month?" Maharaj-ji said, "All right, a year from March." So it just turned out that it was exactly a year from March when the Indian government threw me out. Now you can see no obvious cause and effect between Maharaj-ji's prediction and the action of the Indian government, but once you've begun to see how the game works, you wouldn't trust anybody as far as you could throw them. You don't know who works for whom anymore. It isn't even done on this plane; that's what's so bizarre about it. As the Indian government was about to throw me out, I said, "Maharaj-ji, you promised." I assumed that when I was pure enough I would feel pure. I didn't know what it was going to feel like, but I knew it would feel different than the way I was feeling. So he said, "Here, eat this mango." Well, I've read a lot of holy books so I figured, "This is *the* mango." I took it into the bathroom so I wouldn't have to share it with anybody. I didn't know whether to plant the seed so that I could have more, or whether this would be enough. I ate the mango, and nothing happened. It was just a good mango.

Then I was leaving for America and he said, "I would never let Ram Dass do anything wrong in

45

America." I figured, "O.K. I'll hold him to it." So I came back to America and started to do my thing some more. And slowly the pulls of the world started to get at me. It had been easy to be very focused on God sitting in a temple in India. Particularly that year. There had been a fire ceremony. At the end of the nine-day ceremony, you take whatever you want to get rid of, put it in a coconut shell, and throw it into the fire. I wanted to get rid of my lust. After all, I was forty years old at that point: enough, already! It had been my total preoccupation for thirty years. If I hadn't had enough by then there obviously wasn't enough, so I decided to give it up. So I gave it to the fire.

The next day was the Ram Lila. They were going to burn a huge straw effigy of Ravanna. Ravanna's the bad guy in the *Ramayana*; he was a huge ego and had ten heads, all filled with desire. You could throw into the effigy whatever within you was Ravanna-like. I figured, "Well, I'll be doubly safe and throw my lust into Ravanna." They took the torch and they lit Ravanna—he was sitting on a huge chair—and they lit him right between the legs. Very symbolic. It turned out it was Yom Kippur that night, so to be triply certain I covered that base too.

For three months, it seemed to have worked. But then, I was sitting on a double-decker bus in London, and I noticed my eyes looking down at the sidewalk, following an attractive being down the street. And I thought, "Uh-oh, here we go again." So it was that I returned to the West, ready again to be a spiritual teacher.

I was planning to return to India again in two years: that would have been in 1974, but Maharaj-ji died before that. Now when Maharaj-ji dropped his body, my intellect said to me, "Where could he go?" Because sometimes in the past I would sit with him, and I would see that physical body and then I would quiet down in meditation and I would feel his presence on another plane. And then I would shatter that one, and meet him on yet another plane. My body would start to shake with shakti, from the amount of energy

46

coming from these different planes. I would move through plane after plane of meeting him in different ways.

I remembered the story you've perhaps heard about Ramana Maharshi, who was dying, and his devotees said, "Babaji, please don't leave us. Heal yourself." And he said, "No, this body is used up." And they said, "Don't leave us, don't leave us." And Ramana Maharshi said, "Don't be silly; where could I go?" Which seemed to me to be the most concise statement of the whole illusion of body. But somewhere inside me was a whole different story. I knew that I wasn't cooked; Maharaj-ji was the cook, and he had just left.

When they burned Maharaj-ji's body, different people saw different things at the burning. Most people were crying and wailing and feeling, like I was, that we had lost our Guru. There was one man who stood by the fire, just laughing and singing, singing, "Sri Ram Jai Ram Jai Jai Ram," all through the night. The next day they asked him, "Why were you laughing and singing?" And he said, "Maharaj-ji was sitting up laughing and Ram was pouring ghee, clarified butter, on his head so he'd burn faster; and Brahma, Vishnu and Siva and all the Gods and Goddesses were raining down flowers and everybody was happy." Now was that man deluded or was that reality? One woman saw Maharaj-ji get up on his elbow and wave at her as if to say, "Don't get upset, Ma," and lie back down and get burned.

For two years, then, I had been incorporating what I learned from my Guru, living and teaching as best I could, hoping that his, "I would never let Ram Dass do anything wrong in America," meant that my impurities would not create karma for other beings. I was doing certain things to keep myself as straight as I knew how. I had a Volkswagen camper and I would go off, say, to the desert in Arizona for six weeks of seclusion. I was cleaning up a lot of my game that way. During those years I was taking acid once each year, to find out what I was forgetting, to uncover any subtle ways in which I was conning myself. One year

47

I took it in the Mid-America Motel in Salina, Kansas; that was my mid-America trip, and the last time I took acid was three years ago in the Organ Pipe Monument Park in Arizona. Though I continually felt Maharaj-ji's presence, I still wanted to experience it even stronger because he is my way, and I wanted to get on with it.

Then in the summer of 1974 I was at Naropa Institute teaching a course in the *Bhavagad Gita*, a course for which I felt Maharaj-ji was giving his blessings. There at Naropa I was part of a whole other scene, because Trungpa Rinpoche represents a different lineage. I found myself floundering a little bit because my own tradition was so amorphous compared to the tightness of the Tibetan tradition. Trungpa and I did a few television shows together. We did one about lineages and I felt bankrupt. I had Maharaj-ji's transmission of love and service but I knew nothing about his history. I didn't know how to talk about what came through me in terms of a formal lineage. I was also getting caught in more worldly play, and I felt more and more depressed and hypocritical. So by the end of the summer I decided to return to India. I didn't know what I'd find, but I'd go anyway. I knew I was different than I was ten years before, but I was still not cooked, and what we owe each other is to get cooked.

Driving east, I stopped overnight in Pennsylvania at a motel where I was planning to watch the House Judiciary Committee hearings on television, but a storm put out the electricity. It was too early to go to sleep so there was nothing left to do but meditate. After about fifteen or twenty minutes Maharaj-ji came to me in a vision. He looked just like he always had looked. He laughed, and spoke to me. It's interesting—he spoke only Hindi, and my Hindi was very bad. In India there was always somebody translating. But on these other levels, the transmission is in thought forms, and then it comes out in whatever language you think in. So he said to me, in very good English, "You don't have to go to India. Your teachings will be right here." It was so vivid, and so real, that at that moment I decided not to go to India. I decided to go to New

Hampshire, meditate a month or so in a cabin, clean out my head, and see what would happen next.

On the following day, passing through New York City, I called Hilda Charlton, to say hello. She told me there was a woman in Brooklyn that I should meet. When I resisted because I only wanted to be alone, she told me that this woman said that my Guru was sitting in her basement.

Of course I decided to stay one more night and the next day I went with Hilda to see this lady named Joya. We went down into the basement of her home and there she was sitting in what Hilda said was samadhi. And I checked: I could find no breath or pulse. She was like a rock. She was a very unusual looking woman; she had long false eyelashes, heavy mascara, and a low cut dress. Maharaj-ji was an old man in a blanket, but after all I'd given up having models about what packages the next message is supposed to come in.

Finally she came down, looked at me, and said, "What the fuck do you want?"

Hilda said, "Oh, dear, this is Ram Dass," which didn't seem to make any impression on this lady at all. She said, "I don't care who the hell he is. Does that old man over there belong to you?" I looked, and there was a blanket with nothing on it. So I said, "I don't know." She said, "He's buggin' me; get him the hell out of here."

Then her consciousness shifted just a bit and she went into a very light trance, and suddenly Maharaj-ji seemed to be speaking to me through her. He was talking about things that he and I had been discussing in India when I had seen him last, little matters about maintenance of the temples in India and all kinds of very picayune stuff that she probably could not know and I hadn't even remembered. She came back from that plane but, as she explained, she was not conversant across planes so she didn't know what had just happened.

And I was pleased, because this experience following so closely the vision in the Pennsylvania motel seemed the answer to my prayers.

A few months later I moved to New York City where, for fifteen months, I studied intensively with Joya.

The teachings had a bizarre intensity that it is difficult to convey. From five a.m. until one or two a.m. each day, it was like being caught in a tornado or tossed in a giant clothes dryer. One had to either get out or give up. Surrender was ordained an absolute necessity to experience the higher teachings of this quite unusual teacher. Surrender and devotion was a method I had opened my heart to through the teachings of my Guru, so this process was just a deeper letting go; a letting go of even my resistance to much of what seemed to go against common sense. I also at this time received ghastly reports from various of Joya's closest followers that my resistance was causing Joya bleeding and intense psychic pain. There was no alternative but to suspend all judgment and surrender into these teachings, to simply allow the teaching to come through and burn away all my preconceptions of what a teacher should be or how a teaching should be conveyed. And I let go into what I deeply believed to be a pure transmission. And as I surrendered ever more deeply to those teachings I stated publicly that, as Joya had professed, she was an enlightened being; a statement which I have since come to regret. The intensity of the confrontation (often twenty hours a day) forced my subtle ego defenses to the surface. And Joya in Kali-esque way pounced on these impurities and magnified them until I had to let go or get out. I let go of these impurities as fast as I could and I hung in as best I was able. This was just the fire of purification that my chronic case of unworthiness was seeking.

The intensity of the dramatics and the brilliance of the staging and props created a reality that made me ready to believe the bizarre assertion that a Jewish housewife and mother of three who was married to a very fine Italian Catholic businessman in Brooklyn was in fact Ms. Big, the creative force of the universe. Joya represented herself as an actual form of Kali as well and as a number of other cosmic identities including Athena, Sri Mata Brahma the Mother of the Universe, and Tara the Tibetan Goddess of Tantra. It was a hard act to follow.

Several hundred of us were seduced into this reality by a combination of her powerful charisma, her "chutzpah," and that she seemed often to go into deep trance states with a cessation of bodily functions; too, she reported that she had manifested the stigmata, and certainly knew things which a tenth-grade-educated person would not be expected to know. The stage was well set and we went for it because our greed and our spiritual materialism led us to greatly want to believe it.

In the beginning, Joya spent much time in trance states in which she apparently functioned as a medium. Through her came many seductively-rich teachings from Biblical, Hasidic, Hindu and Buddhist wise men and women of the past, or from beings on other planes. Her voice and language often shifted from unschooled Brooklynese to exquisite poetry that poured forth for hours at a time. I was breathless with the richness of these moments.

I was led more and more to surrender to the reality of the entire scene because we were told that it was only through such total loving surrender by those around her that these higher teachings could come forth. She told me that some of my teachers at that time were such august spiritual figures as Jethro (Moses' father-in-law), Padma Sambhava, Lao Tzu, as well as Ramakrishna, Christ, Mary, Nityananda, an early Kabbalah teacher, Kali and Durga. Having never been around people in trance states this whole scene really astounded me. I was fully seduced by the whole melodrama, like a tourist, open-mouthed, watching a fakir do the Indian rope trick.

Joya kept reiterating that she had come to earth only to be an instrument for my preparation as a world spiritual leader and that ultimately she would sit at my feet. It sounded a little grandiose for oddly enough each day I felt myself more and more becoming just nobody special. There were moments in fact when I felt a bit like Krishnamurti being herded towards leadership of the Order of the Star just before he resigned, leaving fifty thousand members who thought he was the new world leader with a message that they should look within and not seek the Dharma anywhere outside of themselves.

A deep concern in the period just before I met Joya was that I was not yet free of my attachments to sexuality. After a long and intense bisexual history I still found that my perceptions were being colored by my sexual desires. I could afford to be patient about my own purification from sexual clinging, but in view of my public role, I was uneasy that any sexual preoccupations on my part would subtly contaminate those I worked with either in lectures or individually and thus reinforce their own attachments and suffering. Despite the fact that Maharaj-ji said, "I would never let Ram Dass do anything wrong in America," the persistence of these sexual preoccupations led me to question Maharaj-ji's meaning and deeply yearn to clean up my sexual act. In view of how many years I had been trying to get free of these sexual clingings, including offering my lust into the sacrificial fires of India, I had given up hope of ever knowing freedom in this lifetime. The sexual karma just seemed too heavy.

I had read of the Tantric initiations in certain Tibetan sects for just this purpose. The monk would go through a series of ritual openings working with a dakini, a heaven-realm woman. Mostly these were young women who had been prepared from childhood to serve in these rituals without any personal involvement or clinging to the sensual aspect of the ritual. In my fantasies I was hoping that at some point I too would be introduced to such teachings and through such conscious rituals with a disciplined guide become unattached once and for all to these desires.

And now I was presented with a woman teacher, who within a few months after the commencement of the training began to focus on my sexuality. As I opened more and more, assured by her of her perfect non-attachment to any desire system, I felt a new hope that my dream for purification was finally manifesting through this teaching. I plunged headlong into the tornado, casting caution and doubt to the winds.

Perhaps most important of all the considerations effecting my deep involvement with this teaching was that Maharaj-ji had again and again said to me, "See the world as the Mother and you will know God." He often was heard to be repeating the word "Ma" over and over again. He had a shrine built to Durga, an aspect of the Mother. All of this Mother devotion made me feel a bit like an outsider. My own feelings about mothers were colored by the relationship with my own mother and my training as a Freudian therapist and theorist. To be in love with a universal mother just wasn't yet occurring for me. I yearned to understand this aspect of devotion. For I knew that devotion to the Mother, just as devotion to Hanuman, the servant of God for whom I had overwhelming love, was a part of the lineage of my Guru. Sooner or later I felt I would find a way into a devotional relationship with the Mother. When I came to New York City and started to study with Joya and enter her matriarchal reality I felt that at last I had come into the teaching I had sought for so long, particularly when Joya further professed to be the Divine Mother herself.

The fact that Joya continually spoke about Maharaj-ji and implied his presence by seeming to carry on conversations with an astral Maharaj-ji whom I could not see, fed into my longing and somewhat shaky faith that, though Maharaj-ji had left his body, he was still around to guide my spiritual journey.

Joya seemed to have great difficulty staying in her body and would, at the slightest provocation, go stiff as a board. Efforts to keep her in her body, to keep her from just leaving her body behind and going on to other realms, consumed much of our time together. There was a jewel that Joya wore around her neck that Hilda had invested with a mantra to bring her down. When Hilda touched the stone, Joya usually came down but with the pain, so she said, of a thousand razor blades cutting through her. This was in turn very painful to all of us. We therefore went to great lengths to surrender to Joya's every whim so as not to be responsible for this painful drama.

With her increasing feeling of power she also cast aside Hilda. Hilda, while not being a very strong source of teaching, had, as Joya's compatriot, generated with her astral carryings-on the necessary climate of semi-hysteria to sustain Joya's melodrama.

But it was becoming increasingly apparent that what seemed to have begun as a spontaneous mediumistic opening apparently was just too much energy and power for an unprepared individual with power needs and love needs of her own. It seems that the temptation to misuse the trust and power for personal aggrandizement and emotional reinforcement was just too much for her to overcome. Instead of remaining an empty vessel which on occasion contained the wisdom of the ages, Joya claimed the contents as her own, indeed she claimed to no longer be the vessel but to be only the source of the messages which came through—it was a bit like a cup which, filled with seawater, claims to be the ocean itself.

There were just too many "signals" like the moment Joya and I were hanging out and the telephone rang. She picked up the receiver and in a pained whisper said, "I can't talk now, I'm too stiff," and let the receiver drop. Then without hesitation she continued our conversation as if nothing had happened. I realized how many times I had been at the other end of the phone.

And I became bored.

For several months I interpreted my boredom and heretical thoughts as my ego desperately defending against the ultimate surrender.

No matter how I rationalized, my doubts and boredom grew. The tantric exercises no longer seemed productive. I began to experience Joya as another person with attachments. So I began to entertain the possibility that these feelings were cues that I was finished with this teaching and should leave.

Increasingly there came the recognition that though all these planes and beings are simply fascinating, it just isn't the same as liberation. There was more power, more light, more energy, shakti just pouring through us, beings appearing to us, great teachings, wisdom, knowledge; but you notice that your power trips are still there, are still active. You watch yourself think, "Boy, I waited a long time for this." And then there's another plane and another. But it's just another space, and attachment to that space is just more suffering. As the Sixth Zen Patriarch reiterates, "Develop a mind which clings to nought." All these planes are listed in the yogic tracts. But they're all just stuff. They are interesting and useful to loosen your hold on this plane and to transmute and to burn out stuff but ultimately it's just more stuff. Because experiences in meditation and shakti experiences, just like experiences on acid, ultimately must be let go of. If you can give it all up then you just eat your karma alive, you just consume your impurities. Then you can go beyond polarity, beyond pleasure and pain and awaken out of the illusion of your separateness.

And you begin to understand that you have taken birth in order to go through a series of experiences until you transcend the dualism of experiencer and experience. You reside in being, not becoming. Until you can *be* rather than just know the teachings, so that you *are* the teaching.

By the end of this period, I felt I had finished my work with Joya and many of the beings who taught through her. It just wasn't what I needed anymore. It was like trying to find out how many angels can fit on the head of a pin. Finally, the only thing you can do is become an angel and see how many of your friends can hang out on the pin with you.

My doubts grew faster than I could consume them. Joya had changed a great deal in the year. She came to resent having beings speak through her and refused to serve as a medium. Thus while she still had great shakti and charisma, her lectures became merely the reflections of the culture in which she had grown up, sprinkled with spiritual homilies.

As the reality crumbled, I began to see the painful backstage life of the actors and I attempted to bow out as gracefully and as lovingly as possible. Maharaj-ji had warned us that no matter what we did we should never put another person out of our heart. I waited until my love was strong but when I tried to leave it was made very difficult and it became apparent that I was involved in a system which had no escape clause. I had to push against the system though there was very little support for such action. And I began to see the similarity between what I was experiencing and the stories I had heard about other movements, such as Reverend Moon's group, the so-called Jesus Freaks, and the Krishna-consciousness scene. Each seemed a total reality that made involvement a commitment which disallowed change.

My leaving Joya was part of a large exodus of disillusioned followers, including some who had served as servants in her home. As the refugees who left the front lines exchanged stories, the incredible tapestry of falsehood and misuse of the truth started to unravel. It seemed that her incredible energies came not solely from spiritual sources but were enhanced by energizing pills. Her closest confidants now confessed many times they were ordered to call me to report terrible cries they knew to be untrue. They complied because Joya had convinced them that it was for my own good.

Such stories of deception came thick and fast. I had been had.

Since I now see that some of the things I previously said in lectures or articles about the teachings are just not true I come away with egg on my beard. But of more significance than my embarrassment is the issue of truth. In a sense I find myself in a position not unlike that of Mahatma Gandhi when, having initiated a large protest march in which many thousands were involved after the first day's march, he called his lieutenants and cancelled the protest. They objected strongly saying that after all this work and effort he couldn't do this. He answered, "My commitment is to truth, not consistency."

And so I was confronted with the dilemma of how to communicate to the many who had put their deepest trust in me in much the same manner I had put my

56

deep trust in Joya, and how not to let them down in the same manner that I had been so disappointed.

But these teachings had their positive side. Many underwent incredibly deep experiences of who they really were during an intense sadhana they may not have undertaken without the illusion to draw out the energy and commitment needed to do the work that each must do for himself. Through these teachings and the leaving of them many of us have gained more strength and compassion, a greater openness and a deeper ability to allow the moment to be as it is. For all of this I am deeply grateful. However, while I and others profited from these teachings, not everybody did. Some seemed to have been hurt and came away from her teaching with despair, cynicism and paranoia.

So of course the question arises as to whether there is reason to fear taking teachings because the teacher might not be coming from the purest place. I think we need not fear this, for often a student can progress very far, indeed their purification may be greater than their teacher's, because their intention is purer. I got my karmuppance because of my own spiritual materialism. If your longing for God is pure that will be your strength. Then though you may get lost for a time eventually your inner heart will hear what to do and all the impurities in your world will just become grist for the mill.

Lineage

I struggled for a number of years to be a pure eclectic. That is, to be true to each tradition as I was studying it. And yet to be strong enough to be able to hold them all within myself. To contain them all. But what I had to do, which most of you have had to do, is to compartmentalize myself into a series of conceptual groupings so that when you are with the Buddhists, you are Buddha-like; when you are with the Sufis, you are Sufi-like; when you're with the Hindus, you're Hindu-like; when you're with the Christians, you're Christian-like; when you're with the Hasids, you're Hasid-like.

In fact a few years ago we had a retreat at a Benedictine monastery and there was a gathering of the "big boys"; there was Swami Satchidananda, Alan Watts; Sasaki Roshi, Brother David, Pir Vilayat Khan and on and on and on. We each got a chance to do our specialty and everybody participated. So at four a.m. I was sitting next to Swami Satchidananda and we were doing zazen. We were given the koan, "How do you know your Buddha-nature through the sound of a cricket?" Then we were taught how to go in properly with the three bows and the kneeling and the whole proper ritual to get in to see the Roshi. And he sat there with a bell and a stick. It was the first time I had done this. So all the time we were sitting there "being empty" I was of course planning my answer, because I didn't want to make a fool of myself. I mean this was a heavy league to be playing in, you know. So,

"How do you know your Buddha-nature through the sound of a cricket?" and I thought and thought and thought and finally hit on one which I thought would be perfectly appropriate. So I came in and Sasaki Roshi says, "Ah, Doctor, how you know your Buddha-nature through sound of cricket?" And I cupped my hand to my ear like Milarepa listening to the sounds of the universe. I figured I'm a Jewish Hindu in a Catholic monastery so I'll give him a Tibetan answer to a Japanese koan. I was really just delighted with my own cuteness. And he looked at me and rang his bell and said, "Sixty percent!" And at that moment, he absolutely had me. He caught me good in my middle class achievement-oriented identity. We both laughed.

Those moments of connectedness such as with Sasaki Roshi on more than one plane simultaneously are very precious; when you meet another human being both in form and out of form; when you are dancing with them and yet you are free of attachment to the roles in the dance.

I had that moment with Swami Muktananda when he had given me a mantra and in an inner room of his temple the mantra took me up to an astral plane where I met him again on that plane and I looked into his eyes and when I looked in his eyes I started to go up and to fly. As I was flying I started to lose my balance and went to straighten myself out and was immediately brought back to the cave that I was meditating in in his temple. I came staggering out of the cave just having returned to earth and met him out in the hallway and he said to me through his interpreter, "How did you like flying?" And he looked at me with a twinkle. Which was the twinkle of, "You and I just met there; and yet we're here and we're together at all of these places simultaneously." There was that delight, de-light, the delight of that moment.

Alan Watts was a person very much into that space of sharing two planes simultaneously, of that kind of delight. As was Carl Jung who would go up and out into these other realms, these other planes of reality, but when he returned he said, "I was always so happy to get back to my family and to earth, to my home, from these other states." But that's not total freedom, for to be enlightened means you don't cling anywhere; not to this or to that.

And most of the beings I've had this moment with, we could only have a moment. I never knew whether we could only have a moment because there is only the possibility of having a moment or because I or none of us were pure enough to be able to maintain that space continuously. The only person who was always there though I consistently have never been able to locate him in a time-space "fix" was Maharaj-ji. Because no matter where I looked, he was and he wasn't. No matter how high I went he was always sitting there. No matter where I go, I feel him always present. Now that could be an opening of my heart specifically to him. Or it could be something else. Something about him, and his own freedom.

So at the beginning of the journey, you are very eclectic. That's what *Be Here Now* was, it was very eclectic . . . a little of this and a little of that. A Buddhist meditation to quiet the mind. Some Sufi dancing to get the body and heart open. Some Tai Chi when the body's not balanced. Some massage to loosen it all up. A mantra as a centering device. Many methods to choose from the spiritual smorgasborg.

But there comes the point where the pull in you starts to draw you in the direction of what is called one lineage or another. These could also be called ways or aspects of God; they all go to God, but they all come through slightly different routes. They are your dharmic path. So, for one, the route that would be optimal in this life would be to marry and have children, and be a householder, coming to God through your service in that domain. For another person, that would be an adharmic path—that would take them away from God. You cannot conclude that there is any route which is in and of itself the perfect route, for each of you. However, there is a route, and part of the process of tuning is listening to hear how it all is for you. And what you tune to ultimately is the vehicle through which you can sufficiently surrender.

The particular path which is my lineage is what

might be called Devotional Tantra. In addition to my Guru, there are two things in the reality in which I function that are dominant themes. One of them is the Mother, and the other is God. The entire universe, all of its forms, all form is the Mother. You are all part of the Mother. The Mother has many faces. The Virgin Mary, Durga, Lakshmi, Kali. Some are wrathful, some are tender. Some of you are involved in seeing nature as the Mother, that is Mother Nature, but if you expand it outward, all forms become the Mother, and you end up with the interesting choice of either covering the world with the Mother or covering the Mother with the world. For me, the Mother covers the world; for most of you, the world covers the Mother. You are caught in one illusion, and you don't see another that lies behind it.

If you cover the Mother with the world, you are lost in the world and a car is a car, a television set is a television set, anger is anger, doubt is doubt, and a mother and a father are a mother and a father. If you cover the world with the Mother, every experience that you have in this lifetime is but another aspect or face or quality or tone or movement of the Mother and, in this reality, you as a seeker are relating to the universe as the Mother and your relation is one of learning to love, feed from, interact with, consume and finally devour and become the Mother.

I cover the world with the Mother, so that my entire dialogue with the universe, including my own body, is all a form of making love with the Mother, both as a child, feeding from the breast, and as a lover with his beloved, the same way as the gopis with Krishna: it's the same lineage. And it is done through my daily life situations: that's the tantric part of it, through all my senses. My relationship to the Mother is that of the heart; it's the same as that of Ramakrishna to Kali. My body is part of the Mother. I feed on the Mother. I absorb the Mother. I drink the Mother. I feed at the breast of the Mother continuously. The Mother is the shakti, the juice, the vibration, the energy of the universe. I keep growing inside as I feed more and more. I have to consume the Mother.

I consume all of it. I consume the violence, the beauty. I have to even consume Kali, and I have to consume the beauty of the universe as well. I have to take it in, not digest it so it doesn't exist any more, but hold it all in my heart so that I am acknowledging it all.

I am consuming the Mother and I am taking all that energy and using it in order to stay with God. I make love to the Mother. I make love to the universe. In truth, I am the bride of God. For in truth, this game is the game of returning to God. And put another way, in Tantric terms, in true Tantra, not in phony, man-made Tantra, it is intercourse with God. You become both the lingam, the phallus, and the yoni, the vagina; you become both that which thrusts forth into God and that which opens to receive God.

Ultimately, you incorporate it and you open to it, so there is orgasm of the soul, not genital but an orgasm of soul. You keep consuming it all into yourself, and offering it all outward. The flow starts to be so full, it takes you right up to God. Going into God is going into that which is beyond form. Because the concept of God is part of the Mother, of course.

And how do you get on with it? The things that don't get you to God you give up. What do you give up? Unworthiness. You don't analyze it, you just give it up. You give up guilt. Guilt isn't going to get you to God. You give up anger. It's not going to get you there. Do you want to get there, or do you want to screw around? Preoccupation with your own melodrama. Do you want to hold on to it or do you want to get on with it? God doesn't care. You're the one that's in a rush, you're the one that bought this book. You want to get on with it, give it up, it's very simple. That's what Maharaj-ji said to me, "Ram Dass, you're angry?" And I said, "Yeah." I was very righteous and he said, "Give it up." I said, "But . . ." And he said, "Just give it up." And he looked at me like, "I'm your Guru and I just told you to give it up." I'm telling you, if you feel unworthy to be in the presence of God, give it up, don't analyze it, don't say, "I'll plan to give it up," do it now. Here Ram Dass, you take it. Here Kali, you eat it.

Kali eats it. Here Ma, you eat it. Kali eats anger, she eats everything in you which keeps you from God.

Kali is an aspect of the Divine Mother but what a mother to have. She's really gruesome. She scares the hell out of most people. You know why she scares them? Because they want to hold on to who they think they are. She's the fire of purification. She's going to take every single solitary bit of their stash and what will be left are just pure souls floating up into the One. The minute you're no longer attached to your separateness, to your individual differences, to having the universe the way you think it ought to be, suddenly you don't see that form of Kali, you look right through that one and you see the Golden Goddess.

At first, when you have just started that journey back to God, every time you see Kali, who confronts you with all the things you're afraid of—disasters or accidents or you get mugged or you get raped or you lose your job or something "awe-ful" happens—you say, "Oh, get away from me. I want happiness. I want pleasure. I didn't know this was part of the bargain." But later on, as you become more conscious of where the journey is going, you say, "Come on baby, give it to me." That's the moment when you recognize that suffering is grace. And at that moment you become invulnerable because what's anybody going to do to you?

You discover the way in which suffering is the fire of purification; that only when you are lost in your ego do you damn your suffering. When you are a soul yearning to be free, you use your suffering and you use your pleasure. You use it all to get to God, to get liberated. And you begin to notice that your suffering awakens you more than your pleasure. If you seek out the suffering, you are then called a masochist. So you don't seek it out because that wouldn't be honest on the psychological level of reality. But when it comes along, you don't knock it.

"Ah, cancer. Far out." As a being in the body, the temple of my soul for this incarnation, I will do my best to heal it, but I will work with the cancer whether

I am healed or not as a vehicle for awakening. A conscious being uses everything. Nothing goes down the disposal. Including the moment of death, which can be the most profound moment of the incarnation for growth and awakening when you are ready to use it that way. When you don't get so lost in your melodrama, you stop creating more karma for yourself.

Letting go is the act of purification. So all the stuff called the Five Hindrances, or the Ten Fetters: the anger, sloth and torpor, agitation, ill-will, greed, lust—all those obvious ones; and all those subtle ones like attachment to fine material plane things like astral entities—all of this stuff just slows down your journey to God. You finally get so greedy to get done, that you just want to get rid of the stuff. Instead of spending years of analyzing it or therapizing it, like playing with your feces, you just want to get done, you just want to give it all up.

Now, Kali will only come after you if you ask her to, if you don't ask her she won't bother you at all. But if you ask her, she'll confront you with all the "uglies" and then consume your reactions. But if you try to hold on to your reactions, then you're in for it. That's phony holy. If you in truth want to give them up, say, "Here Kali Ma, you take it."

If you don't like the stuff, give it up. You can give it to me, that aspect of Kali that is in me will take it. I won't hold on to it. I'll send it to God, I won't get stuck with your karma, I won't take anything from you unless you give it, because if I take it from you without your wanting to give it, then it's my problem, my karma. And I've got enough without yours.

As you get stronger you begin to experience Kali within you. Then you can consume your own reactions into yourself. And so you consume all of the universe, all of the forms, into yourself. As it gets consumed, the patterns of the forms are converted back into the pure shakti or energy from which they arose. Just keep consuming it, like sucking on a breast, filling and filling. And as you expand out so that the universe is within you, there is nothing that is separate from you, and slowly you get closer and closer, first into the presence of God, and then you start to be fed by God. You actually feel it pouring down into your head.

And any worldly ecstasy you know is as nothing compared to the ecstasy and bliss as your body enters into samadhi, as your breath stops, as your pulse stops, as you sit on the edge of being in God. For then you are beginning to know that part of yourself which is the energy of the universe. All of the sadhana ultimately becomes the preparation of your body, of your heart, of your mind to receive this energy. If your heart is not open and you try to go to God, it will become dry and brittle and you will have much suffering.

When you start to experience so much pressure from shakti, the reason you're experiencing it is because the shakti is out of balance with the love. The way you deal with that pressure is to breathe in and out of your heart. Start to experience the flow, and open up, because you've gotten too dry: Open up to the juices of the universe, and start to make love, and feed from the breast of the Mother. With that opening and flow, you can accept ten times more shakti. You always lead with shakti, follow with love, lead with shakti, follow with love. The extent of it is determined by the depth of your wisdom. They keep balancing each other, until finally they all come together at the end. Because love is the greatest power.

It's the process of going up into samadhi, filling and then coming back down into the physical plane. Ultimately the game is nothing less than this, when you are in what is called sahaj samadhi: every in-breath, you consume the universe, between the in-breath and out-breath, you feed from God, then on the out-breath you feed humanity. On every in-breath you take it all back into yourself, between the in-breath and out-breath, the nectar pours into you, on the out-breath you bring it back down into your incarnation: that is the act of relieving suffering. So that ultimately you become "Not my, but Thy Will be done"; you become the grace of God made manifest; you prepare yourself. You prepare yourself to become an instrument of the transmission of the light. Because, put very simply, without God, life is not worth living. It's as simple as that.

So ultimately each person finds his or her lineage or route through. And when you reach the stage of asking, "God, know me," or "Let me be enlightened," or "I want Nirvana," or however you've said it, at that moment you call forth your spiritual guide or Guru, whom you may not know, and you may never know until the moment of your enlightenment. That being may be Christ, it may be any one of a number of beings, and it is not necessarily on the physical plane. In fact for most of you, your real Guru, your Sat Guru, is not on the physical plane. Your Guru will guide you, to the extent that you are asking purely, through one teaching after another, some of them will be in the form of teachers or situations or experiences. And when you trust that you are in relationship to your Guru, you will constantly learn how to ask your Guru inside, and listen, and tune to the awareness of the presence of your guide, and allow your Guru to guide you, and you will begin to see how each situation is being presented by your Guru to bring you home.

Your Guru or guide represents a unique and specific lineage. Christ represents a lineage. Padma Sambhava represents a lineage. Cochise, the American Indian, represents a lineage. Abraham represents a lineage. These are not lineages that are necessarily identified with any specific religion. Many of the highest beings have incarnated across time and across religion. And the same lineages have come down so that a being could represent a lineage that has manifested within Tibetan Buddhism, within Hinduism, within Judaism, within Christianity. Just as Luke is different from John, is different from Paul is different from Peter; so Milarepa is different from Tilopa, Yellow Cloud is different from Cochise in the American Indian holy man tradition. The different Tzaddiks in the mystic tradition of Judaism represent different lineages. In the Talmud, the different rabbis represent the different lineages. You are ultimately going to make it through on a specific lineage. Or you may not have a guide in form, you might be advait, meaning non-dualistic, the formless, which would attract you ultimately to perhaps Zen Buddhism or Jñana Yoga. And ultimately, you start to fall into a lineage, not because it's the hip thing to do, not because your intellect tells you how it's interesting, not because it's a nice community, and you like the way they dress, but because that way pulled you. It's your way through.

And as you tune to that lineage, your perception shifts, and you begin to notice changes in figure and ground in relationship. You notice teachers you never noticed before, you notice people to be with you never noticed before; the whole process starts to narrow in perceptually, and you start to go directly on what the Theosophists call a "ray" coming from God. Even working devotionally with the concept of God is a ray. For merging into God is merging into where God is not, because it's beyond the concept of God. Where God is not is exactly what the state of Nirvana is. But to know that all ways lead to the end does not nullify the requirement that, sooner or later, you will have to make some sort of commitment or other. A process of surrender is required.

And you go through the lineage. A lineage which is purely defined, in which the teacher is a free being, is one that catapults you ultimately out the other end; it isn't designed to make you a follower of the lineage, it is designed to take you through itself and free you at the other end. A less pure teaching of a lineage traps you in the lineage, makes you a Buddhist or a Christian or a Hindu, not a free being, because when the people that lead do not have the full connection, they cling to the institution, rather than the truth upon which the institution is founded, and institutions corrode unless they are constantly fed by the living spirit. And the living spirit comes only through beings who are it. You can become an organizational groupie as part of your path, but if you know it's not enough, have the honesty to let it go. Ultimately you will come out of a lineage at the other end and acknowledge that through the Sufi, through the Hebrew, through the Christian, through the Buddhist, through the Hindu, through the Zoroastrian, through lineage after lineage, have come beings who have become the living spirit. Then, like Ramakrishna, you will put on each of the hats, not out of need, but out of acknowledgement, to appreciate the universality of ways. A true master, in the perfected sense, is someone who is a statement of the culmination of all ways, even though the form in which he or she manifests may be a vehicle for the transmission of a certain lineage. Ramakrishna, ultimately, was a vehicle for the path of devotion to the Mother. But when he had completed his work, though he remained in the path of devotion to the Mother, he was totally in the advait, non-dual state, way beyond the Mother. So at the beginning is eclecticism, at the end is universality, and in the middle is the lineage.

Guided Meditation

*To be read aloud to a spiritual friend,
as a guided meditation.*

Sit straight, so your head, neck and chest are in a
straight line. Start by focusing in your heart area, in
the middle of your chest where the Hridayam, the
spiritual heart, is located. With your mouth closed
breathe in and out of your chest, focusing on your
heart as if you were breathing in and out through
your heart. Breathe deeply.

Because of the purity of our seeking, many incredibly high beings are present, and with them comes
a great deal of the spiritual substance, out of which
all form derives. You could imagine that substance as a
golden mist which fills the air. With every breath,
don't just breathe in air; imagine you are pulling into
you this golden substance. Fill with it; let it pour
through your entire body.

Breathe in the energy of the universe, the shakti
of the universe. Breathe in the breath of God. Let it fill
your whole body. Each time you breathe out, breathe
out all of the things in you that keep you from knowing your true self, breathe out all of the separateness,
all of the feelings of unworthiness, all the self-pity, all
the attachment to your pain, whether it's physical or
psychological. Breathe out anger and doubt and greed
and lust and confusion. Breathe in God's breath and
breathe out all of the impediments that keep you from

knowing God. Let the breath be the transformation. Now let the golden mist that has poured into your being focus in the middle of your chest, let it take form as a tiny being, the size of a thumb, sitting on a lotus flower right in the middle of your heart. Notice its equanimity, the radiance that makes it bright with a light that comes from within. Use your imagination. And as you look upon this being become aware that it is radiating light. See the light pouring out of its every pore. As you meditate upon it, experience the deep peace that is emanating from this being. Feel as you look upon this being that it is a being of great wisdom. It's sitting quietly, silently, perfectly poised. Feel its compassion and its love. Let yourself be filled with its love.

Now, slowly let that tiny being grow in size until it has filled your body so its head just fills the space of your head, its torso, your torso, its arms, your arms, its legs, your legs. So that now in the skin of your body sits this being, a being of infinite wisdom, a being of the deepest compassion, a being who is bathed in bliss, self-effulgent bliss, a being of light, of perfect tranquility. Let this being in your skin begin to grow in size. Experience yourself growing until your head reaches the top of the ceiling and you are sitting beneath the floor and all of the beings gathered within this room are within your body. All of the sounds, even the sound of my voice, are coming from inside you. Feel your vastness, your peace, your equanimity.

Continue to grow. Your head goes up into the sky, blueness all about, until all of your town, your environment, is within you. Look inside and experience the human condition, see the loneliness, the joy, the caring, the violence, the paranoia, the love of a mother for her child, sickness, fear of death, see it all. Realize that it is all within you. Look upon it with compassion, with caring. At the same moment with equanimity, feeling the light pouring through your being, inward and outward.

Now grow still larger, feel your vastness increasing until your head is among the planets and you are sitting in the middle of this galaxy, the earth lying deep within your belly. All of humankind lies within you. Feel the turmoil and the longing. Feel the beauty. Sit in this universe, silent, huge, peaceful, compassionate, loving. Let all of the creations of human beings' minds be within you; look upon them with compassion.

Continue to grow until not only this galaxy but every galaxy is within you, until everything you can conceive of is within you. All of it inside you. You are the only one. Feel your aloneness, your silence, your peace. No other beings here, all of the planes of consciousness are within you.

You are the Ancient One. Everything that ever was, is, or will be is part of the dance of your being. You are all of the universe and so you have Infinite Wisdom, you appreciate all of the feelings of the universe so you have Infinite Compassion. Let the boundaries of your being disintegrate now and merge yourself into that which is beyond form and sit for a moment in the formless, beyond compassion, beyond love, beyond God . . . Let it all be, perfectly.

Now very gently, very slowly, let the form of the boundaries of your vast being, the One, be reestablished. You are vast, you are silent, all is within you. Come back from beyond the One and slowly come down in size, come down through the universes into this universe, until your head is once again among the planets and the earth is within you. Until your head is once again in the heavens and the cities are within you.

Come down in size until your head is at the top of this room. Stop here for a moment. From this place, look down into the room and find the being who you thought you were when you began this meditation. Look at that being, bringing to bear all of your love and compassion. See the journey of that being as it is living out this incarnation, see its plight, its fears, its doubts, its connection. See all the things it clings to which keep it from being free. See how close it is to

71

knowing who it is. Look within that being and see the purity of its soul.

At this moment reach down from your vast height and very gently, very delicately, with your mind, place your hand very gently on the head of this being, and bestow upon it your blessing, a blessing that in this very life, it may fully know itself. At this moment you are that which blesses and that which is being blessed. Experience both simultaneously.

Come down in size now until you are back into the body which you thought you were when you began. You are still flesh surrounding a being of radiance, of wisdom that comes from being that vast One, of the compassion that comes from being in tune with the truth, and of a love for all things. Feel the love and peace pouring out of you. Use the light that is coming through you now for transmitting that energy, that blessing, to all beings everywhere. Become a lighthouse and send peace and love to all those who suffer.

Think of all the people who you have felt less than love for, look to their souls and surround them with light, with love and peace at this moment. Let go of the anger and the judgment. And then send the light of love and peace out to people who are ill, who are lonely, who are afraid, who have lost their way. Share your blessings, because only when you give can you continue to receive. And you will find that no matter how much you give, you will receive tenfold. As you go on this spiritual journey you must accept the responsibility to share what you receive, for that is part of the harmony of God, that you become an instrument for the manifestation of the will of God.

Now let the radiant perfect being once again assume its diminutive form, the size of a thumb, sitting upon a lotus flower in your heart, in your spiritual heart in the middle of your chest, radiant with light, peaceful, immensely compassionate. This being is love, this being is wisdom. This is the inner Guru, this is the being within you who always knows. This is the being who you meet through your deeper and deeper intuition when you've gone beyond your mind. This is the being who is the flow of the universe, the tiny form of the entire universe that exists within you. At any time you need only sit and quiet your mind and you will hear this being guiding you home. When you have finished this journey, you will have disappeared into this being, surrendered, merged and then you will recognize that God, the Guru and self are one.

Questions and Answers

I really want to be a good yogi and I am trying very hard to purify myself, but it's so hard. What's wrong?

Purification is the act of letting go. There is a statement that appears in one of the Gospels which states that men need not disfigure their faces in order to know God. There is a type of righteousness and seriousness that creeps in the minute you decide you're going to do spiritual practices. Where suddenly it's serious work and you have to be a certain kind of way. Sort of "tight-assed." You may find that though that looks good from the outside, it begins to feel kind of lousy from the inside. And there is a way in which denying too much stops flow. There are a lot of people who are really good meditators, who sit perfectly and their mind gets very quiet. But they aren't liberated. Because they have pushed away form, they've pushed away the earth, they've pushed away the heart, they've pushed away flow.

If I understand this game at all, it's a game of exquisite balancing. And the balancing can be understood within different systems. For example, in southern Buddhist meditation, Theravada Buddhist meditation, there are three components that are emphasized. One is called sila, one is called samadhi, and one is called pañña. Sila is the purification: non-killing, non-stealing, non-lying, right speech, right livelihood and so on. Samadhi is concentration and mindfulness. And pañña is right understanding and right thought, or the wisdom connected with it. Now if you watch the way the game works with those three components—purification, concentration and wisdom—you'll see that you wouldn't even start this dance without a little bit of wisdom. You have to understand a little bit of what the game is about to even want to sit and meditate. So you have a little bit of pañña, and then you try to do samadhi, concentration. But every time you try to concentrate, all of your other desires, all of your other connections and clingings to the world keep pulling on you all the time. So you have to clean up your game a little bit; that's called sila, purification. You clean up your game a little bit and then your meditation gets a little deeper. As your meditation gets a little deeper, you are quieter and you are able to see more of the universe so that wisdom gets deeper and you understand more. The deeper pañña makes it easier to let go of some of the attachments so it makes it easier to increase the sila. And the increased sila allows the samadhi to get deeper. So, you begin to see the way these three things all keep interweaving with one another. They're a beautiful balancing act.

Now in the same way, in my particular lineage, there are balancings that can be understood in other ways. One way is to talk about the heart and the mind. That is, flow and quietness. Another balance is to talk about form and formless. Another way of saying that is to talk about the Mother and the Father. Still another way of talking about a balance is talking about shakti and love or power and flow. Some people get symptoms of shakti—pressure in the head, shaking, movements, twitching, nausea, pains in the back, all kinds of symptoms because of the lack of the balance. Because you're too much into the shakti realm without the flow. But you begin to be able to diagnose your own predicament in your own body, in your own being, to ascertain what is out of balance and come back into the flow. Because getting the powers, the yogic powers, the siddhis, without the love, without the flow, without the compassion, makes you just another power tripper. Our society is full of people who have siddhis, who have powers, they have power of the intellect and they have power of the mind to control others. But the compassion isn't there. The flow isn't there.

On the other hand, people who have flow, lots of loving flow, but no power, no shakti, no control, no discipline, no one-pointedness, tend to get very mushy. They're like soft earth, there's nothing firm in them, and that lack of firmness keeps allowing them only to go so far and they keep falling down again. There's no backbone in the process.

The way to approach the whole sadhana is with firmness and with lightness. Not hysterical 'ha,ha,ha' but with a lightness, a delight, enjoying the light of it all, making it light. Somewhere I remember a line that goes, "The angels can fly 'cause they take themselves lightly." People tend to get very lost in their own melodrama, romanticizing their own spiritual journey. "I'm going to God." And they tend to take themselves very seriously. They've got themselves a story line. And they begin to look like yogis and they begin to smell like yogis, and they come on as they imagine yogis should, they have a whole image of themselves becoming yogis. That's all going to have to go. You come back into the present moment. You are what you are. You let go of the romantic story line of your own predicament. Because that one's just keeping you from being wherever you really are at the moment. It's a self-consciousness.

I'd just like to point out that righteousness, being very "good," does not necessarily bring you to truth. Once you are wedded to and immersed in truth, in the sense of formlessness, then you will be righteous. It's like the Ten Commandments. You can do them out of a "goody-goody" place with anger in your heart, and righteousness and fear of punishment or you can come into the space of your own being in relation to God where you look and see why things are the way they are; and it just flows out of you and you just can't act in ways that create karma or lay trips on other human beings. Then you begin to understand the Ten Commandments from a different angle. Righteousness coming out of truth; rather than truth coming out of righteousness.

My son has developed a neurotic habit which worries my wife and I quite a bit; what should we do?

To the extent that you are free of the attachment to how it ought to be with him, and your identity as a father, still being able to be a father perfectly—I'm not talking about abrogating responsibility for safety and survival, just not getting lost in fatherness—you can see him as a being who is living out a certain incarnation in which this neurotic pattern is showing. By contacting that being behind the neurotic pattern he can drop it when he's ready to drop it.

My understanding of the way in which a child grows is that you create the garden, you don't grow the flower. You can merely fertilize the earth and keep it soft and moist, and then the flower grows as best it can. You create a space with your consciousness that determines whether that neurotic pattern gets deeper into that child or whether it's seen as something that can be cast off when the time comes. If you define this being as "my child who has this habit" and that's the major reality of the relationship between you and the child, that's catching him in the habit. The minute you see him as a soul who's incarnated in this situation in which he's working through this stuff, he's free to drop it whenever he needs to because you're not attached to his having it or not having it.

It's an interesting one, because people get guilty that they're not doing enough about their children and they tend to get caught in this sort of predicament. You don't change your wife or your child. You just keep working on yourself until you are such a clear mirror reflection, such a supportive rock of love for all those beings that everybody is free to give up their stuff when they want to give it up. Your wife, her anxiety; your child, that habit. You keep creating a space in which people can grow when they're ready to grow.

The predicament is that a child and a parent may be at very different levels of evolution in terms of their age of being. A child could be much older than the parent, or much younger than the parent, in an evolutionary sense. There are many old beings being born into this culture at this moment. They have been looking to take birth in a conscious environment. So that some of you have babies that don't particularly want to be incarnated because they're almost beyond it. They're just doing some little clean-up operation.

The minute you do a "take" of beings as souls rather than personalities and bodies, you don't cling to the incarnation that hard. You understand its function and you don't demand that the incarnation be other than it is. You understand that births are consciously chosen to work out specific karmic necessities and you don't get so lost in the melodrama on this plane.

It's very tricky which level of reality you climb into. The power of a conscious being is that they don't use one against the other. They keep all those levels of consciousness going simultaneously. So that somebody is brought in to see me on a stretcher and they're in terrible pain and they have been for years and at one level I can see that they're doing a tremendous amount of work in this life; and at another, "God, this person is suffering so badly, can I do anything to relieve the suffering?" Both of those at the same moment consciously. And if that person who is brought to me is somebody who says, "I wish to awaken during this lifetime; Ram Dass, help me." I say to them, "Well, you're really feeling sorry for yourself. You've really got a good birth; you're cleaning up a lot of stuff. Let's work on how to convert pain." And if they are somebody who didn't say that, but we just happen to meet, like somebody's aunt or something like that, I say, "God, it's really rough how much you're suffering. Here let me fix the pillow for you," or "Are you having proper medical treatment?" or "What can I do for you?"

It's very interesting how you deal with problems and suffering depending upon which plane of consciousness is the dominant theme, although you never forget the other one. A strong consciousness keeps it all going at the same time. You do everything you can to help your son feel more loved, calm, supported and get rid of the neurotic habits—at the same moment you are not attached and you understand that is the karma of this being, being lived out and you work on yourself until you are a perfect environment for that being to do what it needs to do.

How do you interpret dreams?

In general, I'm inclined to suggest you don't do to much analytic work in this dance, because your mind plays too many tricks. If the dream has an immediate significance that affects you emotionally, work with it. It may give you a click into place of something you needed to understand about yourself. Fine. But if you say, "I wonder what that meant," forget it! It fits under the category of things that when you're ready to know, you'll know. Don't sit and analyze or wonder or get preoccupied with it. It all has meaning. It's all work you're doing on other planes. It is significant spiritually, but you don't have to understand it.

You are existing at many planes simultaneously at this moment. The only reason you don't know of your other identities is because you're so attached to this one. But this one or that one; don't get lost, don't stick anywhere, it's all just more stuff. Go for broke, awake totally.

You say every life situation is a perfect lesson; how is that so?

The universe is made up of experiences that are designed to burn out your reactivity, which is your attachment, your clinging, to pain, to pleasure, to fear, to all of it. And as long as there is a place where you're vulnerable, the universe will find a way to confront you with it. That's the way the dance is designed. In truth, there are millions and millions of stimuli that you are not even noticing, that go by, in every plane of existence, all the time. The reason you don't notice them is because there's no attachment to them in you. Your desires affect your perception.

Each of us is living in our own universe, created out of our projected attachments. That's what it means when it says, "You create your own universe." You are creating that universe out of your attachments, which can be out of your avoidances and your fears. So more and more you keep consuming your own reactivity and saying, "Right, and this situation too, and this one too, Tat Twam Asi, and that also, and that also, and that also." Then it starts to lose its pull and fall away.

And you get so that you're perfectly willing to do whatever you do. And you do it perfectly. It's like Mahatma Gandhi gets put in jail and they give him a lice-infested uniform and tell him to clean the latrines, and it's a whole mess. And he walks up to the head of the guards and he says, in total truth, "Thank you." He's not putting them on or uplevelling them. He's saying, "There's a teaching here, and I'm getting it; thank you." What's bizarre is that you get to the point where somebody lays a heavy trip on you and you get caught, and then you see through your *catch-ness* and you say, "Thank you." You may not say it aloud because it's too cute. But you feel 'thank you.' People come up and are violent or angry or write nasty letters or whatever they do to express their frustration or anger or competition, and all I can say is thanks.

When an oppressor, or an oppressive economic system, is causing people suffering, it seems to me not enough to love the individual who is oppressing along with those who are being oppressed. It seems that if one was to really love them, one would speak out. I fear that many of us who seek, feel that it is no longer necessary to criticize in such a manner.

He is raising the question about our social responsibility for political inequities, social inequities ... where is our responsibility? Is it enough to meditate? Is it enough to become a loving person?

Well, our predicament is this: We are in an incarnation. We can't make-believe we're not. We must honor the attendant responsibilities that go along with that incarnation, we must honor parents, political identities, social identities and work, in form, in order to alleviate the suffering at whatever level we find it.

Now, the peculiar predicament is that when you see any kind of injustice in the world, if you are attached to anger about it, or are attached to it being any other way, you are at one level perpetuating the polarization even as you are working to end it.

In the Patanjali Ashtanga Yoga, it says there will be no giving and no receiving. Does that mean that nobody gives and nobody receives? No. It means that when you give, you are not attached to being the giver. Thus, you do not force the other person to be the receiver. The political inequities are our political inequities. There are no "them" in the universe. There is only "us," more or less pure. And we as a collectivity must purify ourselves. Each individual must hear her or his dharma; that is, the way in which the manifestation must come forth in order to relieve suffering. Until you are enlightened, all action must be an exercise in working on your own consciousness. The forms, however, will differ. For example, if somebody comes up to me as my friend Wavy Gravy did last year and tells me that it only costs ten cents a day to keep a person alive in Biafra; his having come forth and told me that creates a situation in which I now exist. That situation elicits from me a set of behaviors to do what I can do—so I do a benefit to raise money to help and have that money go to help feed starving Biafran children. If Wavy had not said that to me, I would probably not have done that.

You can't walk away from life on this plane. For instance, I feel it is dharmic for me to be involved in politics at this time. I supported Jerry Brown when he ran for President, because I felt he was the more conscious being. And I would like to see consciousness involved in government.

There are a thousand and one ways in which man is injust to fellow man. Which will you work to change? Which is your particular dharmic path? And as you work to alleviate suffering, will you be careful that the way in which you do it doesn't create more suffering in the long run? Be conscious. Since you're not fully enlightened, everything you do must be done as work on yourself. At the same moment, you must listen to hear what form your efforts must take to relieve suffering.

You may run a nursery, you may just help an elderly lady across the street, you may go and fight in Israel, you may go into the Peace Corps, you may join a community service, you may go to Washington and work actively in politics, you may work in a free health clinic, you may become a concert guitarist. You may raise your children with great love and consciousness. We are not in the position of judging each other. Each person must hear his or her own dharmic way. What you feel is most important may not be seen as most important to someone else. This is a very complex society we are a part of. Stay in the world, do your part, raise your children, earn your living, assume your responsibility at every level. Do it all as an exercise to bring you to God. For until you are with God, every act you perform will both liberate and entrap. And if you are really interested in ending suffering, you recognize that the end of suffering is full awareness. And a person can only help another become aware when that person is aware.

It's only because we forget the First Commandment in the first place that we're dealing with all this right now. So now we're in the process of remembering. It's very simple. Can you hear that?

What's the best way to deal with the judging mind?

Watch it. Watch it do its thing. There it is judging again. Very simple. If you're working with the Mother, you can offer it up to the Mother. If you're working with Vipassana meditation, you would merely take your primary object of meditation, which might be following the breath, and then every time a judgmental thought came up, you'd note, "judgmental thought" or something like that, and then you'd go right back to the breath. It's just another thought.

Thoughts keep clothing themselves in all kinds of silk and glitter and they say, "I'm not just another thought . . . I'm *you*." You know. "I'm real. This judgment is the *real* thought." But it's just another thought. This whole game is just thought.

82

After a really good meditation, I feel like I'm not in my body.

You probably aren't; it's true. I must honestly admit that I am of the school of hard knocks. I'm not going to protect you from confronting all of your attachments. The reason you're not in your body is because you don't want to come back into your body. You're attached to the high. Okay. Confront it. If you're aware of it enough to complain to me, you're seeing your own predicament. I think that we are not brought through by a spoon-fed operation. Maharaj-ji would allow me to enter absorption states where my body would be shaking and the breath would become all but non-existent then he'd say to the interpreter, "Ask Ram Dass how much money Steven makes." I'd struggle to ignore him but he'd demand I come back immediately. You learn after awhile that you have control, that you can do all this stuff. There's no real need to protect you from yourself. You're just seeing your own attachments.

Why did it all begin? Why did we leave God in the first place?

That is the question which is the ultimate question and Buddha's answer to that question was, "It's none of your business." Which is not a facetious answer. He's saying your subject-object mind can't know the answer to that question. It's an answer which you can be but you can't know, because in order to know that you would have to be that from which it started, but you aren't as long as you're asking the question. It's one of those kind of absurdities that you get caught in. There are a dozen different answers all of which are equally real and unreal. You might say God took form in order to know himself, that he had to become separate in order to see himself. Or it could be said that since there is no time at another level of reality, nothing happened anyway. That's a real answer too. These are all valid answers within one level of reality or another. Every level has its own answer to that question but in truth the answer is not knowable

83

until you have transcended those levels because every answer you give is just feeding your mind from one level to another and they're all only relatively true. Now that all sounds like words which means it's not an appropriate question. You keep asking it but you won't get an answer. I mean, not only from me, you just won't get an answer.

What is shakti?

Shakti is the universal stuff from which it all comes. Everything here is shakti, it's all just shakti, patterns of shakti. It's the stuff of the universe, finer than quantas of energy in the physical, scientific realm. You can ignore it if your method does not involve focusing on energy or you can work with it, draw it in, mobilize it, direct it and use it as a force. You can use it in the same way you might use electricity, you can collect it in the same way, it feels the same way except it is much finer. You can draw it in and draw it in and draw it in and you will experience new realms of perception and new powers.

Before you are done, you will be subject to, or must surrender to, intensities of energy that grow and grow and grow, until they are nothing short of all of the energy of the universe, and to the extent that there are impurities within you, or paranoia, or a body that is not kept in a good shape, when you start to tune in on these higher energies, you can really blow your circuits, or shake yourself very badly. When you see people shaking, all that bouncing energy stuff, that doesn't have to be. That is because the person is trying to put 220 volts into a 110 volt system. The process of purification is preparing yourself as a container to handle more and more energy, more and more love, and for that you need a quieter and quieter mind and a stronger body, and a more open heart.

There are a lot of different traditions, some of which are very much oriented around shakti; Kundalini Yoga for instance. Others recognize and use these energies in another manner. When you just work with shakti, you get great power. But unless that shakti is perfectly balanced with wisdom, and the empty mind, and love, it can be extremely destructive. Similarly, if you only work with your intellect and with the emptying of your mind, as in some yogas, and you fail to open the heart, your journey becomes very dry and brittle. And though you enter God, you don't enter in fullness. By fullness, I mean in complete flow, an open fluid relation to the universe. Ultimately no matter what your method you have to get a very even balance between your energy, your heart and your mind.

How does LSD affect the spiritual journey?

My first struggle with that was in a correspondence with Meher Baba back in 1965, in which he said that very few people can use it positively; for many people it will make them insane. And I wrote, "It's strange, Meher Baba, but the only reason I read your books is because I took acid, and that's true of many devotees that follow you in America." And he wrote back and said, "I know you're a good person, and for a few people it can be helpful; but for most it's not helpful; and you can take it three more times." Well, I didn't listen to Meher Baba; I took it a number of times more than that. Then in 1967 or '68, Maharaj-ji asked me about that medicine that I used in the West and he took 900 micrograms, as you may know from *Be Here Now.* Nothing happened at all to him, which was impressive. I must admit though that because nothing happened I went through a little doubt. I thought maybe he threw them over his shoulder, maybe they never got in his mouth, it all happened so fast and when you're around somebody like that you're so stoned, who knows. So I had this little doubt, but I came back and told everybody he took 900 micrograms. In 1970, when I was in India the next time, he said, "Ram Dass, did you give me some medicine the last time you were in India?" "Yes," I said. "Did I take it," he said with a little twinkle in his eye. I said, "Well, I think so." He said, "What happened?" And I said, "Nothing, Maharaj-ji." And he said, "Jao! Jao! Go away." The next morning he said, "Do you have

85

any more of that?" So I brought out what I had left and he took twelve hundred micrograms this time. He took each tablet and struck it in his mouth and made sure that I saw, and he munched them up. Then he said, "Can I have water?" I said "Yes," and he asked, "Will the medicine make me insane?" So I said, "Probably." So he said, "How long will it take?" I said, "An hour at the most." So he got an old man up with a watch and he was holding it and looking at it. And he drank a lot of water and about half way through, he started to look really weird, he even went under his blanket, and he came up looking totally insane. "Oh my God," I thought, "what have I done to this sweet old man? He probably threw it over his shoulder last time, and he wanted to show me what a big man he was." At the end of an hour, he looked at me and he said, "You got anything stronger?" Because nothing had happened, obviously. Then he said, "These were known in the Kulu Valley long ago, but most yogis had forgotten them." On later questioning, he said, "Well it could be useful, in a cool place, where you are feeling much peace, and your mind is much turned toward God, and when you're alone." He said that it would allow you to come in and pranam or bow to Christ, but you could only stay for two hours, and then you would have to leave again. He said, "You know, it would be much better to become Christ than to just visit him. But your medicine won't do that, because it's a false samadhi"—which was exactly what Meher Baba had said to me. "Though," he said, "it's useful to visit a saint; it strengthens your faith." Then he added, "But love is a stronger medicine."

Now, what I did was, after that, once each year I would take LSD when I was peaceful and was alone and my mind was turned toward God, to sort of find out what was happening, and each time was a very profound experience. These past few years, when I started to do this more intensive work, there came a point in my sadhana where I was working with extraordinarily intense shakti energies, and I came to be aware of an incredible amount of tension in the back of my head. This turned out to be the result of the immense amount of acid that I had taken which, while it did not permanently physiologically damage me in any way, had created a kind of psychic resistance; psychic in the same sense that chakras are psychic phenomena, rather than physical phenomena. The usage of acid had created a psychic blockage in my medulla, which made my spiritual work, for quite a while, more difficult, in terms of working with these incredible pressures in my head. It turns out that had I not ingested so much psychedelics I would have been able to get through that space much faster. So my conclusion about it now is: For those who don't know about other levels of reality it could, under proper conditions, if they are truly turned towards spiritual life, show them that possibility. It did so for me. Once they know of the possibility, and they really want to get on with it, the game is not just to get high again, but to "be" and *be* includes high and low.

It is also true now that the culture has shifted and different kinds of realities are now accepted in everyday life. There are many young people who never took acid and never smoked grass whose consciousness are free to float, in the midst of this culture, in and out of planes, and that's partly because of the Rock and Roll movement; and, too, it's partly a result of the cultural shift which resulted from the usage of acid in the sixties. Don't underestimate the social changes that have occurred as a result of the acid phenomena.

I don't deem that for a being on the spiritual path the LSD experience is any longer necessary. It is very clearly not a full sadhana; it won't liberate you. Because there is a subtle way in which there is attachment, in the sense of feeding your unworthiness because you aren't it without it; and you have to look outside yourself to get hold of it. As a method it also has the limitation that it overrides stuff that you would best deal with. Grabbing at experiences and pushing aside old habit-patterns in order to get high is ultimately just delaying the process. Because ultimately you have to confront those habit-patterns and purify them.

After you know of the possibility you get on with it and any time you're just after another experience you're just getting more hooked on experiences and experiences are all traps. The game is to use a method and then when you're finished with it to let it go. This isn't a good and evil matter; it's just a question of honesty with yourself as to whether in fact you are using your opportunities in order to awaken as effectively as you can.

How do you open your heart?

A good exercise is to do deep breathing in and out of the heart as though it had nostrils, right in and out of the heart. You can use that breath to ferret out those places in you where there is a deep sadness or some deep attachments which are slowing your progress; let them come forth and let them go, give them up to Kali or Christ or Guru or God. Keep bringing them out, the sadnesses deep within your heart that have closed you off, keep bringing them forward, keep going in and in until you're all the way back to your spine. Keep digging and digging with the breath, breathing it all out again.

Another way is to go out into the woods or to the ocean to concentratedly make the physical gestures of opening the heart space, like Hanuman does. Physically, though of course you don't actually pull apart your flesh, you open your heart and you call upon whomever you're in close contact with, perhaps Christ; you might say, "Christ, let me feel your love." You're not asking him to love you, you're asking to be allowed to feel the love he has for you. If you really open yourself and ask that in truth, you will feel a warmth starting to touch you which will permeate you and start a process of your opening. Or you can sit with a picture of some being like Christ and just experience that love flowing back and forth between you and the picture in deep breathing.

It is just so incredibly gentle and beautiful starting a dialogue of love with a being who is love. Some of you have known Meher Baba who is such great love, or Christ who is a statement of love, who is love itself. You just open yourself. You sit in your little meditation area and you take a picture of a being whose love is pure, whose love is in the light of God. It's not the love of personality, it's not the love of romance, it's not that needful love, "I need you." Romantic love is jealous and possessive because the object of that relationship becomes your connection to that place in you which is love. The kind of love that Christ gives is conscious, unconditional love, he just is love. And ultimately you become that kind of love. There's no need about it. You're living in that space and don't need anybody to turn you on to love because you are it and everybody that comes near you drinks of it.

And as you become more and more the statement of love, you fall in love with everyone. There's nobody here that I don't feel in love with when I look at because all I see is that which is love in them. I can see all their impediments too, it's all there but I'm not climbing into it. I'm not denying it, but I'm not getting stuck in it either. And because you feel love when you are with me, that opens you to the place in yourself which is love. Sometimes when you feel that, you want to cling to me because I'm your love connection; but I'm not clingable to. There's no way you can collect me; the only way to do it is to become it yourself. Otherwise you're always going to be looking for connections. Most people want a Guru because they want a lover or a father. In fact, the Guru can be the guide to the beyond. Don't listen to what other people say about the Guru or even to what the Guru says about the Guru, listen to what your heart says about the Guru.

If you follow your heart nothing will happen to you, you are protected. As long as your actions are based on your pure seeking for God, you are safe. And any time you are unsure or frightened about your situation, there's a beautiful and very powerful mantra—"The power of God is within me. The grace of God surrounds me"—which you can repeat to yourself. It will protect you. Experience the power of it, it's like a solid steel shaft that goes from above through the top of your head right down to the base of your being. Grace will surround you like a force field. Through an open heart one hears the universe.

How do you interpret statements like, "No man comes to the Father, but through me."?

In almost all holy books, and especially in the words of holy beings, you are dealing with transmissions to different levels of disciples and devotees who can hear different things. In the *New Testament* are these the words of Jesus or of the Christ? You have really at least two beings in that one being. One of them is Jesus who is the Son, a form of the Father made manifest on Earth: "I am in the Father; the Father is in me." Then there is the Christ, which is the consciousness out of which that form is manifested, the consciousness that acknowledges the Living Spirit. That's not necessarily Jesus, the man. The predicament, depending upon your degree of readiness, is that you either become involved with the devotional relationship to Jesus, the man, or to Christ the consciousness; Christ, the love. And my experience of that particular biblical quote is that it is Christ speaking, not Jesus; that Jesus is an historical statement of the perfection made manifest, and at that historical moment Christ said to somebody, "You can only come to the Father through me," though it may have been interpreted as coming from that body which was Jesus. For someone else, at another moment, it means the greater body out of which that body comes, which is the Christ body. And that Christ consciousness is what would be called the Living Spirit. It's like the statement, "Eat of my flesh; drink of my blood." He didn't expect people to come up and tear off his arms or drink out of his veins; that is the universal form speaking, saying, "Consume the universe into yourself; drink of the universe so that you may know the Father." That's not Jesus speaking, that's the Christ.

And the problem occurs that much heavy violence has been done interpreting that initial statement as a statement of Jesus rather than as a statement of the Christ. Its misinterpretation has led to proselytizing which has led to a lack of acknowledgement of other people's ways of meeting the Christ other than through the form of Jesus.

A standard criticism of not only the spiritual practices but all forms of religion is that it's an opiate of the masses, it's a way of escape, it's a tool of the ruling class to take people's minds off of the social struggle and on some pie-in-the-sky that they think will solve their problems, but won't really. What about it?

That's a very complex issue. In one sense they're absolutely right in that when you enter into these other realities the social/psychological/economic preoccupations and hardships look entirely different. I have met beings who have living conditions that I would consider subhuman, who are totally radiant, luminous, fulfilled, happy beings. Nobody's exploiting them, this is the way they *are*, they have choices, but it just doesn't matter to them. I look at them and I don't see somebody who is drugged in the sense of "the opiate of the masses," I don't see somebody who has lost their freedom. That being has found something that makes worldly concerns less relevant to them. That doesn't make them bad or good, or weaker or stronger.

However, if spiritual seeking is used by one group of people to control another, that's another matter. Nobody has the right to control the consciousness of another human being. That applies to revolutionaries as well as the establishment, and if I choose to sit quietly and be totally fulfilled in a room with no furniture, living on bread and water, I don't think I have to define myself as underprivileged or as suffering because I live below the standard of living. If somebody laid it on me against my will, that's oppression; but if I choose it as a means to extricate myself from deep conditioning that's my business. Don't let paranoia rule the game about who's doing what to whom.

Jerry Rubin just came back from China and he said to me, "What are you using terms like 'God' for?

They're anachronisms, you're just taking people backward. China's got the greatest religion of all time because it's right here in the world."

And I said, "Well, Jerry, that's still a limited view of what human beings are about, and they're still suffering in it." It just depends on the level from which you view the dance and the freedom of individuals to choose the way they want to live. I believe in external *and* internal freedom, and I won't surrender my internal freedom for external freedom. Most Western activists want freedoms they can see and measure, the external freedoms. But somebody who is seeing clearly, I think, can recognize that even when you get all the external freedoms, which many people in the society actually have, you are still not free. That's what spirituality addresses itself to, the matter of inner, or internal freedom. Once you have internal freedom, you may or may not be a political activist, you may or may not be an artist, you may or may not be anything. Most likely, you won't sit around apathetically. But there isn't any rule that says you can't. That's external freedom. To say that everybody who is more conscious must be politically active is naive, as far as I'm concerned, because a society is an extremely complex and exquisite organism, and it takes all kinds of parts to make it beautiful.

I see the evolutionary political change as very exciting, like a Martian takeover, rather than everyone picking up a gun and starting to shoot each other. It doesn't have to be "we" against "them." It's we become them and then them becomes us. But it's scary, because there are no symbols to hide behind. Some who come and hang out with me are lawyers, doctors and college professors. I don't say to them, "Give up being a doctor," I don't tell Jerry Rubin to stop being an activist, or Jerry Brown to stop being the governor, or John Lennon to stop singing. Just do whatever you're doing in a way that increases the connection of humanity, the awareness of the interrelatedness of all things. That includes ecological sophistication, and economic and political awareness. Stay doing what you're doing, because there's no one role that defines the game. Just because you march on City Hall doesn't make you an effective political activist. Christ and Buddha were both effective political activists each in his own way. I think we have to acknowledge that there are a variety of strategies in this game of life. And it isn't good guys and bad guys, it's just individual difference.

For those people who find themselves in a particular time and place in which it is appropriate for them to struggle politically against oppression and injustice, could spiritual practices help them?

Yes, because in terms of the effectiveness of any action, you are more effective when you are capable of being totally involved in what you're doing, and totally non-attached. Though I understand how the term "non-attached" might seem antagonistic to the original concerns which have motivated the involvement. Let me elucidate. Part of the total coolness that is needed when under stress comes from compassion for the entire predicament. That is, having an overview of the whole game board . . . it's like fighting a ground battle but you have the additional perspective of a helicopter overhead studying the entire strategy. It allows you to not get so lost in your emotions that you make the other guy have to stay polarized. In other words, you give space for him to grow by seeing how he got caught in his predicament. For instance, I can understand Nelson Rockefeller's position though I don't agree with it. I can protest his actions and say he shouldn't have done certain things and argue for his punishment for the responsibility for Attica, which I do. But at the same moment I can hear his predicament. And that ability to hear his predicament gives him an opportunity to grow. Because every human being has the right to get unstuck from his or her models. But the minute you take away their opportunity to grow, even if they're a bad guy, you've imposed on them exactly the wrongs you'd like to right. Don't create polarization in your zeal to override the bad guy, you might create more. As was said in *Be*

Here Now, the hippies were creating the police and the police were creating the hippies, it was so obvious in the Haight-Ashbury. The citizens got frightened by the scene, so they demanded their police get more oppressive, the police got more oppressive, and that became a symbol against which the hippies mobilized to fight, the more the hippies mobilized to fight, the more the police got oppressive; each force was creating the other. And nobody in that space was conscious enough to cut through that polarization, which could have turned it into a whole collaborative dance together. I think that one thing Jerry Rubin and I have agreed about is that spiritual awareness and compassion and consciousness contribute to political effectiveness.

You know Allen Ginsburg was incredible in the 1968 Chicago Democratic Convention. He just goes and chants OM right in the middle of the scene. Now that's a very interesting mixed game. At the time I was sitting in a temple in India. I read some clippings about Allen in Chicago. And I went through a few changes, like, "Am I copping out? I mean, here's my buddy right there being maced and beaten. What am I doing? I'm sitting in a temple in the Himalayas in this room huddled in a blanket making tea for myself. Is this a cop-out or am I confronting other subtle demons for all of us, which in a way is as difficult as the demons, the bad guys, of the external physical plane. What can I bring to my fellow man?"

And it turns out that I do have something to offer Jerry Rubin, or Jerry Brown, or Allen Ginsburg, or John and Yoko, or perhaps anyone who might share these words. It's like I honor each person for his contribution, but recognize that at the same moment he is reaching for something that'll make his flow more free. Sitting in that temple did actualize something of use to relieve suffering on the physical plane.

It seems that a lot of revolutionary tactics in this country have won the battle but lost the war. If you alleviate human suffering on one level but your act doesn't allow it to be alleviated at another level, then you haven't accomplished the goal of ending suffering. Like in getting economic benefits for people, if you

94

deepen their attachment to thinking that economic benefits are going to give them total peace or happiness, you are perpetuating the illusion that causes the suffering. That's why the nature of the consciousness of the revolutionary determines whether the revolution ultimately liberates or entraps those it was meant to aid. It's a really beautiful issue. Really it's like the Europeans who originally came to America and thought that if they got political and religious freedom, they would have it made. Well, they came here and they got it and they didn't have it made.

How does one decide to get rid of sexual desires? I'd like to give them up but I don't know how.

We have finally found out in America that there really is nothing wrong with sex. We don't have to be Victorian about it. We've gone from neurotic sex to reasonably healthy sex and that's really good. And if you're living in the world, sex is a very beautiful part of existence. However, if you in truth want to go to God in this lifetime, then you start to direct your energies towards getting there. The predicament with sexuality is that no matter how nice your intentions are, the act itself is so powerful that it catches you into the gratification that comes from your separateness, which means sensual gratification. And in that sense it's reinforcing your separateness. You don't give up sex because it's bad or wrong; no guilt, nothing like that. What you don't do is *give up* sex. What you do is acknowledge how much you want God and turn your heart and mind in that direction without sounding like coming on for Barnum and Bailey. You can't get into a struggle against it. Because every time you're busy struggling against something, you're reinforcing its reality. The game is just to go into the reality where sex is like rubbing sticks together to make a fire. You get to the point where you're already existing in that place you were having sex to get to.

I used to insist that in my advanced classes everybody be Brahmacharya, celibate; which usually makes the classes small. But even when I didn't, couples would say to me, "What's happening? As I get more into the trip, our sex life means less and less to us. Something's wrong. Isn't sex divine?" Yes, sex is divine, but the reasons for having sex were falling away. Later they could have sex without any opposition or drain on their inner work. High beings can have the most incredibly beautiful sex imaginable because their hearts are open, which most people having sex in this culture don't experience. The problem is that most high beings don't have any desire for orgasm because they are already in that place.

The issue of Tantra is often played with by people who desire sexual gratification. And they try to have their cake and eat it too. But in truth, when you desire to have a sexual relationship with another person, the arousal process and the gratification is reinforcing that desire. The only kind of truly Tantric sexuality that is possible is between two human beings that are so rooted in God that there is no preoccupying desire for the other person as 'other.' Then, you may use the physiological process of body interaction in order to awaken energy to move it up through the chakras. But that is only when there is no preoccupying desire whatsoever in either partner. And that is a condition that hardly anybody I've ever known could fulfill. Short of that, just be honest with yourself; sexuality is sexuality, it's not Tantra. The true Tantra is basically the relationship between Radha and Krishna, between the seeker and the Mother where you open your soul and become both the lingam and the yoni, both the phallus and the vagina. You are both penetrating into the universe, and drawing it into yourself. Because the soul is neither male nor female. And when you have identified yourself as a soul going to God, the sexual dance starts to lose its pull.

But now I must caution you in view of what I'm saying. Each of us is at a different stage in our evolutionary cycle. Many of you have much work to do in interpersonal relations, sexual gratification and so on. You are at the stage where you want to want God but you have other business to attend to first. To make believe you are done with something you are not done with will slow you down in your journey to God. To try to hold on to something you are done with will equally slow your journey to God. There is no simple rule of the game of who becomes Brahmacharya and who doesn't. Some people do it, and some don't. Married couples may be Brahmacharya or they may not. Brahmacharya couples have sex in order to produce a child, and that's it. Not once a month on the new moon and all that jazz.

You don't discard your balls or your vagina in order to go to God. It's all part of the dance. And just like ultimately you can eat whatever you want, ultimately you can do whatever you want. This isn't a moral issue at all. So that if you can hear and be honest with yourself, you'll know when you are done and when you are not done and when one desire system is stronger than another. Just be straight with yourself. Don't make believe. Phoniness is the worst part of spiritual life. People trying to be something. While some of my students I encourage to be Brahmacharya, others I encourage to have sex. There are many horny celibates in this world that are not going anywhere except to psychiatrists. And there are a lot of people balling who wish they weren't anymore, but can't stop because they think they ought to be. They've already entered into planes of consciousness where it's irrelevant. Trust yourself, allow your desires to be shed when it's appropriate.

What part does diet play in spiritual work?

As I hear it about diet, at different stages of your sadhana, different diets are indicated. You start to be pulled towards them. These are not based on morality. They are based on what kinds of vibratory rates you can ingest and transmute. And there are stages where you can't handle meat because of the vibratory rate, the rajasic, active quality of it, the hot intense passion of the stuff. You can't get calm enough through it. So your diet starts to lighten up, to fish and eggs and vegetables and grains, dairy products and fruit. When you can't handle that pretty soon you might get down to grains and dairy products, vegetables and fruit; then there are times when you can't handle anything but fruit. And then you may go through a stage where you are so connected and clear and beyond it you can eat anything again.

There are certain diets that will help purify your system when it's full of toxins from the kind of crap we usually eat all the time. They're really generally good. A macrobiotic diet will clean you up. So will simple vegetarian diets. But don't get into a good and evil trip about it. It doesn't work. You're just getting caught in a lot of righteous morality. A lot of people are more preoccupied with what goes in their mouth than what comes out of it. I must honestly tell you that people have gotten liberated eating anything. So, it's clearly not going to be simply that game. The American Indians consumed buffaloes and there were some very high mystics and saints among the American Indians. The Tibetans ate meat and they honored the animal and it was all part of the karmic working through for all of them. And the way they did it was not spoiling or wasteful or angry or anything. It was in the way of things.

I remember sitting meditating in Big Sur in a house that was loaned to me by Esalen Institute which came with a cat. Every morning the cat would go out and get its prey to eat. And it would come in, and because it loved me it would come over to me and sit between my legs as I meditated, and chew on the skull of a mouse or a lizard which would sometimes still be alive and flapping. And I wouldn't know who to hate or who to love, or what to do. I learned a great deal. I was taken through a tremendous understanding of one level of our existence.

My diet has recently been a modified vegetarian; that is, I eat fish and eggs. And I do that because, at the stage I'm at, with the shakti I have to work with, my body needs a certain kind of protein which I have to feed it. But many students I require to be either strict vegetarians or yet more lax. Just as I do, you too must listen to your own needs.

*As a woman psychotherapist I'm having a difficult
time integrating what you teach with my daily work
with my patients. Could you reflect a bit on this
condition?*

I think that the polarization of inner work and
outer social action is a polarity that merely comes out
of an attachment to a model in one's head. From my
point of view, both of those come together very much
in Karma Yoga, the yoga of daily life: Doing the daily
actions of life in such a way as to come to a clearer
state of consciousness or deeper peace or enlightenment or whatever metaphor you wish to use. The
work you're doing becomes your practice rather than
taking you away from your daily life. That is, if you
just start from where you are, not where you wish you
were, and your givens are certain trainings and skills
and responsibilities—then the game is to find within
all that the path to enlightenment and how to use it all
as a method of working on yourself.

Now I spend almost all of my time serving, being
available to people who are suffering in one way or
another. It's hard to define who's suffering how, who's
suffering more than another. When people come to me,
my interaction with them from their point of view is
allowing them to re-perceive their life strategies and
their emotions and so on, but really they are my work
on myself, just as arduous psychiatric patients are a
psychiatrist's work on himself. If you get lost in pity
or anger or rejection or desire—sexual desire or desire
for power over your patients—then you become less
effective as an agent of change. Part of your work is
to deal with counter-transference and your own emotional reactions to people. From my point of view, my
work is to stay in a place of total involvement in the
psychological plane with total non-attachment. I do
what I do, and I do it as perfectly as my consciousness
allows it to be done, except that I'm not attached to
how it comes out. I'm just doing it as best I can. It
comes out as God wants it to come out, not as I think
it ought to come out. That is, when I meet people, I

don't immediately know they ought to change just because they are in a mental hospital. I have no reason to say that how I think they ought to be is better than how they are. I just share my being with other beings, and they change to the extent they are capable and ready and can change, using my consciousness as an instrument.

Your struggle happens because your model of being a psychiatrist or being a woman or being any label is entrapping, because labels are limiting, they are finite, they have suffering connected with them. And part of the work of consciousness is to redefine your own being, your own nature, to the point where you *are*. Then there's psychiatristness and there's womanness and there's personalityness and there's opportunityness and so on; these are like phenomenal rings around your essence rather than who at center you really are. As long as you think you're somebody who's doing something, you're lost in the illusion and can't really offer anyone else the space to extricate themselves from their negative reality.

Now the optimal strategy in behavior change, with yourself and every other human being, is compassion. That means, as far as I understand it, the ability to see how it all is. As long as you have certain desires about how you think it ought to be, you can't hear how it is. As long as I want something, I can't really understand it, because much of what I can see are just my own projective systems. You must come to see every human being including yourself as an incarnation in a body or a personality, going through a certain life experience which is functional. You allow it to be just the way it is at the moment, seeing even your own confusion and conflict and suffering as functional, rather than as dysfunctional.

The greatest thing you can do for any other being is provide the unconditional love which comes from making contact with that place in them which is beyond conditions, which is just pure consciousness, pure essence. That is, once we acknowledge each other as existing, just being here, just being, then each of us is free to change optimally. If I can just love you because here we are, then you are free to grow as you need to grow, because none of it's going to change my feeling of love.

We're used to having these special role relationships, thinking certain roles apply to one, yet not to another, because we're very attached to externals—do you touch somebody, do you sleep with them, do you beat them, do you control them, do you collaborate with them, do you support them, do you pay them, do they pay you? That's all stuff of the vehicle of the interaction between two beings; it isn't the essence of the matter. As you work on yourself through your daily life, more and more you see your own reactions to things around you as sort of mechanical rip-offs. You get much calmer in the space behind it all, and you're able to hear more how it all is, including your own personality as a part of nature. The deeper you are in that space, the more there is available for everybody you meet who is capable of coming into that space. You are the environment that allows them to do that—and from within this space all change is possible. The minute you identify yourself or anyone else with models, roles, or any characteristic, any individual difference, change is really fierce. Once you live in a universe where you experience even your living and dying as relative rather than in absolute terms, it's all free to change. There's nowhere you have to go to work on yourself other than where you are at this moment, and everything that's happening to you is part of your work on yourself.

Different ones of us are different parts of the corpus of civilization, and no one act is any better than any other—if you didn't have the shoemaker, we would go unshod; you need the shoemaker. Is the shoemaker better than the psychiatrist, or worse? And what about the garbage collector? Without garbage collection, you know where New York would be? or Boston? So, is the garbage man more important than the psychiatrist, or less important? The whole thing becomes absurd. You begin to see that everybody, even the president, is just another instrument in the dance, another part of the total body, and each of us must hear what his particular or her particular route through is, and not try to define, "That's a good one and the others are bad," or "That's the best one," or "I'm doing the most important work." The most important work you can do is the perfect job for *you* to do. Discover how to serve people not out of the fact that you're supposed to or *ought* to, just do psychotherapy because that's where you're at. Do therapy as long as you realize that here we are behind doctorness and patientness; here we are behind neurosis and relative neurosis.

A very lovely psychiatrist in Boston is also interested in meditation and the spirit. It's far out, because his teacher at the moment is a man by the name of Karmu in Cambridge. Karmu is a seventyish Black automobile mechanic who, from eight in the morning until six at night, works on cars, and then comes home to his apartment, sits and drinks white wine and red wine, and all these kids come and hang out around him because he's a wise man. And he heals them. He pounds on their bodies, but all the time he's transmitting this incredible unconditional love, because he's loving the place in them behind all their crap and all the stuff and all the divine dance. And here's a psychiatrist who's willing to sit at the feet of this automobile mechanic because he's knowledgeable enough to respect wisdom.

I have three major instructions for my life from my Guru: Love, Serve, and Remember. Love everyone, serve or feed everyone, remember God. My own yoga seems to be doing every day whatever it is that I do—being with people, sharing time with them in whatever way. I don't demand they call themselves patients. We may meet under any circumstances, in a restaurant or somewhere, maybe a bus, and be with one another in whatever way we need to be with one another. In all cases it's my work on myself, because I am loving, serving, and remembering, but what I love and serve is a function of what I remember. What I remember is who we all are. I remember the Self— and that remembering means that my love and service towards another being are directed towards the place in them in which they are already free.

A few years ago I used to meditate and I felt wonderful. Then my life changed and now I look back and wonder what happened to that beautiful state of mind I had.

One difficulty most of us have is interpreting our suffering, and our doubt, and our confusion, and our loss of faith as part of the process of awakening. We keep feeling we fell out of grace, we blew it. "Why aren't I high? What happened? Life stinks. Before it used to be all sweetness and light; and now it's so heavy for me." Not for all of you at all moments; but every one of you has those moments. I sure do.

It's like when Christ comes forth and performs all these miracles and says, "Look, it isn't the way you think it is at all. You aren't who you think you are, I'm not who you think I am, we are all in the Father. Come on, wake up. Let go of all your worldly nonsense. Let's get on with it." And everybody around him gets hooked on him because he's got all this power. Then he leaves them and everybody gets depressed. They got hooked on their method, their method of getting high, and the method split. If you're a drugee, you ran out of drugs. Or like for some of us, our Guru left his body. Or a method that's been getting you high for years—singing to Krishna or following your breath—suddenly turns to straw in your mouth. It doesn't work any more. What about all those lows? When you're angry. When you're getting fired. When you've run out of welfare. When your car breaks down. When there's an unexpected pregnancy. When there is a fight. When there is violence in the neighborhood. When there is racial tension in the community. When there is ecological disaster imminent at every turn. When there is politics that has a ring of fraudulence about it.

And all of this does an interesting thing: it throws us back in upon ourselves, for us to see where we're at.

103

When all the pins get pulled away, you have a chance for a moment to see what resources you have. There are many stages on this path, many lessons; don't stop anywhere. It's all part of the process of awakening. You have all the time in the world but don't waste a moment.

What is the Hanuman Foundation which you initiated and which sponsors you?

The Hanuman Foundation is a tax-exempt non-profit corporation that sponsors at present a number of projects. One is to create a broad base of meditation in America. In some countries, such as Burma for example, people on their vacations instead of going to the seaside go to meditation centers. Soon we too will have more social institutions for meditation. We will soon be publishing *The Game of Awakening: a beginner's guide to meditation,* a book on meditation facilities and strategies presently available in this country.

Hanuman Foundation is also involved in converting the prisons of America into ashrams. The Prison-Ashram Project encourages inmates to redefine the opportunities for inner growth available in prison and what the prison experience has to offer. For, if you've ever noticed, prisons are like monasteries. They give you a cell and they give you clothes and food. They maintain your outer needs to free you to do the inner work. Can you imagine a cellblock with fifteen inmates and three guards in which the inmates all think they are in an ashram, and then the only people left in prison are the guards. So what we are doing is an infiltration of models from within, and to this end we are publishing *Inside Out,* a free book to give the inmates much of what he needs to know for converting his experience into a profitable one for becoming aware, for coming to God.

Another project of the Hanuman Foundation is support for the shifting of metaphors as to what death is in this culture. For death is primarily controlled by the medical professional whose commitment is to preserve life and sees death as their enemy. But we wish to help the dying to use death as an experience for awakening. And ultimately, perhaps, we will have an ashram/monastery/center for dying where people will come to meditate and work on themselves. And part of their work will be to work with people who have come there to die consciously.

An additional project of the Hanuman Foundation is to introduce the concept of Hanuman to America, who is a perfect model of being in God through service. And because of the purity of the love of God, Hanuman has the power to do anything. He is the manifestation of Christ's statement, "Had ye but faith, ye could move mountains."

Also, we are involved in bringing over to this country teachers who embody the living spirit of other traditions.

The other thing that the Hanuman Foundation does is make tapes of my lectures and retreats, and a few other teachers' lectures, available at cost. For more information on the foundation, or if it feels right to contribute to the economics of this work, you may contact the Hanuman Foundation at Box 928, Soquel, CA 95073.

It's a very far out dance to stay in form and know that there is nothing to do but be with God. Then you are free to do all of this stuff so easily.

Who do you think you are? How do you view your role in the contemporary scene; how do you relate to it?

The most honest answer is, I have absolutely no idea who I am. I am not even fascinated any more with who I am. I used to be fascinated with Ram Dass. "Wow, look at that, isn't that interesting?" I try to get a rush off it now and then but it's just not happening. Last week John and Yoko came to visit me. And a few days later so did Jerry Brown. The old mind thought, "My God, a governor has come to my motel room, I must be somebody." And I try to get a little leverage off it, but it's empty, it's just nothing. I'm an old power tripper, so why aren't I getting rushes off power? I can talk to your questions and we can play this thing out. But most of the time I'm just sitting pretty empty. It's quite incredible. It's the consciousness of other beings that draws all this stuff out of me. When I'm with you and you ask me who I am I can give you a nice erudite answer. I might say I'm trained to be a wise man in this society. And I think that's a role that the society sure as hell can use. Now to get really far out, in a previous incarnation I apparently was somebody called Sammat Guru Ram Dass. He lived in 1600 and was the Guru to King Shivaji. And he lived in a little mud hut next to the palace. At one point the king was a total devotee of the Guru and came out with a scroll offering the whole kingdom and gave it to the Guru saying, "Here, I'm giving you the kingdom." The Guru said, "Fine," and gave it back to him, "Now you run it for me." So I've already done that trip.

My Guru said to me, "Lincoln was a good president because he knew that Christ was president and that he was only acting president." I don't really have any personal identity with my game at all because I don't think it's me doing it. I am a perfect robot in a way, a robot for the flow of the universe. I can't think of anything nicer than to be an instrument of the flow of the universe. I'm perfectly happy to be a pebble on the bottom of the lake. That's what's fun about having fame when you don't want it. There's a freedom, not an entrapment, in it. There isn't much anxiety in my game any more, because I'm just as interested in it whichever way it goes.

Twelve years ago I was really a bad guy in this society, an acid head thrown out of Harvard and all that. Now I'm a good guy in the society. Maybe I'll be a bad guy in the society again someday. It's just the flow, just the dance. I just notice it.

All I really want to do is be with God. And more and more, no matter what I'm doing, that's all I'm doing anyway. I'm sitting here and it's just going through me; it's just nothing. It's totally beautiful and the more nothing it is, the more beautiful it gets.

Dying:
An Opportunity for
Awakening

Some years ago as my mother who was dying of cancer got closer to death I spent quite a bit of time with her. Because I was starting to be into meditation and using psychedelics, I found that I was not particularly anxious about this death, even though my love was very strong. As I would sit in the hospital, often high on one chemical or another, I would watch the parade of people coming into my mother's room—the doctors, the nurses, my relatives, my father—all with a total bravado: "You're looking better today," "Did you eat your soup?" "You'll be up and around in a few days," "The doctor says there is a new medicine," and then they would go out in the hall and say, "She won't last a week."

I saw that she was surrounded by a ring of complete, well-meaning hypocrisy, but all I would do is sit next to her very quietly and sometimes just hold her hand for long periods. One day in the darkening room about a week before she died she said to me, "You know, I know I'm going to die. I wish I had jumped out a window when I had the strength to do it." Then she said, "There's nobody else I can talk to about it but you." Until then, she and I had never had a conversation about any of this. She said, "What do you think death is?"

I said, "Well, I don't know, but when I look at you it's like I see somebody I love, a very dear friend, inside a building that's burning. I see the building being destroyed but somehow, as you and I talk, you and I are still here and I have the suspicion that even though the body is being consumed by this illness, not much is going to happen." We stayed there in the silence together, just holding hands, for many hours.

It may be relevant to relate two things about the funeral. For forty-four years my mother and father on their anniversary had exchanged, along with gifts, one rose that was a token of their love for one another. At the temple the casket was covered with a blanket of roses. As the casket was wheeled out of the temple, it came by the first pew. In the first pew were seated my father, at the time a mid-sixties, Boston Republican lawyer, ex-president of a railroad, very conservative; my oldest brother, a stockbroker-lawyer; my middle brother, also a lawyer but one who was having spiritual experiences; me; and my sisters-in-law. As the coffin went by the first pew, one rose fell from the blanket of roses at the feet of my father. All of us in the pew looked at the rose. We all knew the story about the exchange of one rose, but nobody of course said anything. As we left the pew, my father picked up the rose and was holding it as we sat in the limousine. Finally, my brother said, "She sent you a last message," and everybody in the car at that moment agreed. Everybody said, "Yes!" The emotions of the moment sanctioned an acceptance of a reality totally alien to at least three members of the group.

Of course, then the question was, how would we preserve the rose? It turned out that my uncle knew someone who had a process where you could put a flower in a liquid and encase it in glass and it would last forever. So the rose was put in this container and placed on the mantelpiece as the final materialistic hold on my mother. A few years later, after a proper period of mourning, my father took a new wife, a very lovely woman. The process of preservation of the rose, however, hadn't been quite perfected and the color had gone out of the rose into the water, so now it was a ball full of brackish water in which there was a dead flower. There was my mother's message sitting there. It was put in the cabinet in the back of the garage for memorabilia that we can't bear to part with. I'm sure you all have a place like that.

The other part of the funeral that I was going to tell you about was that I took LSD to go to the funeral and it was quite interesting, because I experienced Mother and I hanging out watching the whole scene. She didn't seem particularly upset, nor was I. The family was seated on one side and the rest of the people on the other, so that they could watch the family mourn. I was at the end of the row. It was a sunny day and there was a golden light coming out of the casket. Mother and I were hanging out watching all the people who were thinking beautiful thoughts about her. I wanted to smile, but I realized that would be the final straw: "Of course, he takes drugs, he smiles at his own mother's funeral!" A smile is not currently an acceptable social response at the time of somebody's death.

A few years later, prior to meeting Maharaj-ji, I had visited the city of Benares, also called Varanasi or Kashi. It is a very sacred city in India. At the time, I knew nothing about Hindu religion or mythology. On the streets in Benares were beings who were just at the point of death, many were lepers. They would sit in long lines with their begging bowls around the burning ghat. The burning ghat is a platform that goes out into the Ganges River, where all day and all night one entire caste of people just keep the fires going in which bodies are cremated. When a person dies in the vicinity, the body is brought through the streets on a stretcher wrapped in cloth, either carried by chanting men or put in a bicycle rickshaw. When the body is taken from the home, it is taken with the head toward the home and the feet towards the Ganges. The recitation on the way is, "Ram Nam Satya Hey, Ram Nam Satya Hey. . . ." Halfway to the ghat a ceremony is performed and the body is turned around so that the head is towards the Ganges, because the home is now beyond.

Each of the nearly-dead beggars on the streets had attached to his loincloth a little bag, which I learned contained enough money for the wood needed for his funeral pyre. At that time the poverty in India had deeply frightened me. Their predicament seemed so horrendous to me that I could hardly bear it.

Five months following, when I returned to Benares after having lived in a temple and begun to understand what Benares was about, I walked through the streets and saw an entirely different scene. Because, as it turns out, Benares is one of the most sacred cities in India and to die in Benares is the highest desire of the truly spiritual Indian. It assures liberation, it's a way in which you consciously go toward your death. When you are on your funeral pyre being burned, Shiva, one of the forms of God, whispers the name of Ram, another form of God, in your ear and you are liberated. So these beings who looked to me so horrible were the ones that were succeeding out of all the millions of people in India, these were the ones that had gotten to Benares, that were going to die in Benares. When I looked at them with this knowledge, what I saw on their faces as they looked at me was pity. They were looking at me with pity because I was this foreigner who would probably never die in Benares. I was just caught in the wheel of illusion and suffering, going on and on, while they had made it. I did a complete about-face and began to see that Benares was a city of incredible joy, *even though* it held incredible physical suffering.

So here I am in this forty-five year old decaying body, it's the package in which I am functioning; I honor it, I take pretty good care of it. Yet, whatever the catalyst, whether it was Maharaj-ji, or psychedelics, or my studies, or meditative experiences, the importance of my body and personality, of the Ram Dass melodrama, have been appreciably lessening. Simultaneously, my anxiety about death has concommitantly been dissipating and this new perspective has allowed me to reflect upon death and to write about it. Of course, I was deeply influenced by studies of the *Tibetan Book of the Dead,* by Aldous Huxley's description of dying in the book *Island* and by his own death, and by my mother's death as well. It was apparent that we needed new ways of looking at death.

A number of other experiences contributed further to my appreciation of the matter of death, including two Maharaj-ji stories that affected me strongly:

One day Maharaj-ji was walking with a devotee of many years and Maharaj-ji suddenly looked up and said, ". . . . . Ma just died." She lived in a distant city and it was obvious he had seen this on another plane. Then he laughed and laughed. His devotee was shocked and called Maharaj-ji a "butcher" to laugh at the death of such a beautiful and pure woman. Maharaj-ji turned and said, "What would you have me do, act like one of the puppets?"

Another time as Maharaj-ji was sitting with a group of devotees, he suddenly looked around and said, "Somebody's coming," but nobody heard anyone. A few minutes later an employee of one of Maharaj-ji's devotees arrived. Before he could say anything, Maharaj-ji yelled, "Yes I know he's dying, but I won't come." The man was shaken by these words because his employer had just a few minutes before suffered a severe heart attack and had sent this man running to fetch Maharaj-ji to his side. But no matter how the employee or others pressed Maharaj-ji, he refused to go. Finally Maharaj-ji took a banana and gave it to the employee and said, "Here take this to him. He'll be alright." Of course the man rushed back with the banana, and the anxious family mashed it up and fed it to the sick man. Just as he finished the banana, he died.

Here in America one day Wavy Gravy called me and said, "There is a fellow dying who would be interested in visiting with you. He's at Tom Wolfe's house." At that time I was staying at Baker Roshi's house at the Zen Center in San Francisco. I was alone, since the Bakers were in Japan, and rather depressed. However, I couldn't very well refuse, so I went. The boy was in his twenties, extremely thin, dressed in Levi pants and jacket, and boots. I sat down with him and said, "I hear you are going to die soon."

111

"Yeah," he replied. So I asked him if he wanted to talk about it and he said okay. So we started to talk and after about twenty minutes, when he went to light a cigarette, I noticed that his hand was shaking and it hadn't been before. Because of the paranoia that had accompanied my depression, I thought, "Oh, wow, look what I'm doing. What right do I have to be coming on to him? He's the one that's dying." So I said to him, "Hey, man, I'm really sorry. I don't mean to come on to you, I'll leave you alone. I didn't mean to bug you."

He said, "Oh, no, you're not! I'm nervous because I'm so excited about being with you. I've been looking for the strength to die and you are the first person I have been around who isn't making it worse by being so freaked about it." He was giving me the legitimacy that I didn't have myself, he was saying, "It's okay to do what you're doing."

So we started to hang out together and at one point I rented a car and we went for a ride on Highway One in Marin County, a very perilous highway along the coast. We stopped for gas and he said, "Would you mind if I drive? It will probably be the last time." Now that's a pretty romantic image if you know how twenty-three year olds are about driving, so I said, "Sure." He started to drive but it very quickly became apparent that he was too weak to turn the steering wheel. We were only going about twenty miles an hour, but he would come around one of the curves where there was a 300 foot drop to the ocean and I would hold the wheel and turn it, trying to do it unobtrusively so that there would be no social blunder; I didn't want to upset him. As if the situation weren't bad enough at that point, he tried to light a cigarette. I thought, "Wow, he's not only going to go, but he's going to take me with him!" Then I saw that I had entered into a conspiracy, I was joining with him to make believe he was other than he was, in order to protect his image of himself as this young, virile, twenty-three year old. I said to him, "You know, you've got me caught in a conspiracy, because obviously you can't turn the wheel, let alone smoke at the same time.

I should be driving. We should be into where it *is*, not where we wish it were. This is the way it is now, let's get here." That started a new dialogue between us that got much more exciting, and we just lived in the present moment more and more. The only preparation for death, it turns out, is the moment-to-moment life process. When you live in the present now, and then this present, and then *this* present, then when the moment of death comes you are not living in the future or in the past. The freaky thing about death is the anticipatory fear. But you can't tell someone else to live in the present moment unless you are.

Later I was invited to work with a lawyer who had cancer and I said, "I only work with people if they want to work with me." I was assured, "Oh, yes, he wants to work with you." So I went out to his place, which was a very posh house by the ocean. He was sitting there surrounded by his family and friends, and they were all drinking. He was ruling the whole scene with an iron hand, because he was the one that was dying. I came out and they offered me a drink and I sat looking at the ocean for a while. I heard hysteria in the conversation, the kind in which everyone is laughing too loudly. Finally, I turned to him and said, "I understand you are going to die soon." If he had not invited me to come to deal with death, I would have had no license to say a thing like that. You can't go up to somebody and say, "I hear you're going to die." You have no license to lay your trip on somebody else, but he had invited me to do what I was doing, it was the compassionate thing to do. The whole place freaked. I had said the thing that nobody ever says, and it entirely changed the space. Then the family and friends and I and he all got into a discussion of death sitting by the ocean that beautiful day. We meditated on the ocean and the whole thing started to take on the power of the immediacy and exquisiteness of the oceanic dying process.

Somewhat later a very dear friend was dying in Los Angeles and I went to visit her. She was a very subtle, intellectual, liberal, sensitive person. The first time I saw her, she was at the stage where she was

113

interested in death and wanted to talk about it. As I talked I could hear that intellectual place in her that had not had any experience with reincarnation saying, "I'll listen to it, but it's hogwash." I saw my words were just not doing it. The next time I went to visit her, she was so weak that all I could do was sit by her bedside.

I'll just describe the moment to you. She was dying of cancer of the nervous system and the pain was very severe in her lower abdomen and groin. While I was sitting there she was literally writhing in pain, turning her head and rubbing her hands over her body. Her expression was one of very intense pain. I was sitting next to her doing the Buddhist meditation on the decaying body. This is a formal meditation that one does on the stages of decay of the human body. I was just sitting there, wide-open, not closing my eyes and going off to some other place, just staying with it, noticing the pain, noticing the whole thing, letting my emotions flow but not clinging, not holding or getting into a judgment about it. Just noticing the laws of the universe unfold, which is not easy to do around death because of our emotional attachment, especially with someone we personally love. As I sat there watching the pain and suffering, I started to experience a great, deep calm. The room became luminous. And at that moment, right in the midst of her writhing, she turned to me and whispered, "I feel so peaceful." Though her body was writhing with pain, in this meditative environment she had been able to move beyond the pain and experience deep peace. I, or we, had created this vibrational space that we could be in together. She and I wouldn't have preferred to be in any other place in the universe at that moment. It was bliss.

At a seminar on death and dying guided by Elisabeth Kubler-Ross, a twenty-eight year old nurse and mother of four dying of cancer who had been through eleven operations, asked those of us in attendance, "How would you feel if you came into a hospital room to visit a twenty-eight year old mother dying of cancer?" The answers called out from the

114

audience included: angry, frustrated, pity, sadness, horror, confusion, etc. Then she asked us, "How would you feel if you were that twenty-eight year old mother and everyone who came to visit felt those feelings?" Suddenly it was apparent to all of us how we surround such a being with our reactions to death and forget that there is a being just like us in that body who needs to make a straight contact with someone. It is not unlike the previous point made about the beautiful girl who nobody can relate to other than as a body.

Eric Kast, who did work with LSD in terminal cancer patients, reported on a nurse who was dying of cancer. She took LSD and it was reported in the *World Medical News* that she said, "I know that I'm dying of this disease, but look at the beauty of the universe." Even though she was busy dying she also had for a moment transcended the dying to come into this other place.

A few years ago, a gal named Debbie Mathieson died in Mount Sinai Hospital in New York. Debbie was affiliated with the Zen Center in New York and when she was dying the brother monks and students decided that instead of meditating at the zendo they would come to her hospital room to meditate every night. The first night the doctors arrived on their rounds and pushed open the door with their jovial, "How you doin' tonight?" There were all these beings in black sitting deep in meditation, and the doctors were taken aback. As the nights passed, the doctors would come into the room as if coming into a temple, which indeed it was, right in the middle of Mount Sinai Hospital. In the midst of the hospital, Temple of Life, there it was, a temple to that which is beyond life and death.

In Japan when a person is dying, a screen is placed at the end of the bed which shows the Pure Land of the Buddha and he or she can focus on that so that as death occurs the last thoughts are a reaching out. It's like a railway ticket, it's the ticket that is going to take you through, and you can go out on it if you wish.

These various experiences have led me more and

115

more to the idea of a Center for Dying. This is not new with me, but came from Aldous really. It would be a place where people could come to die consciously, surrounded by other beings who were not freaked by death. I thought it should be near the mountains or the ocean, which obviously have eternality connected with them. Perhaps it would have bungalows and a person could die there in whatever metaphor he or she wanted: as a Christian or a Jew or a Hindu or an Atheist. Those coming to die consciously could die with as much or as little medical aid as they wanted, that would be up to them. While they couldn't ask the doctor to kill them, they needn't have their lives prolonged. What would be added, in addition to the priest and medical staff, would be a guide to help the individual remain conscious and in the present.

Recently, as I have been reflecting on the term "guide," it strikes me that there are no professional diers. Everyone who was dying with whom I shared time helped me, probably more than I helped them. Nowadays many people contact me and say, "Do you think I could work with somebody who is dying?" And I see that from each individual's point of view, in terms of his or her own growth, one of the most profound experiences any of us can have is to work with the dying process, whether our own or someone else's. It confronts us with a number of issues in ourselves that are important in our own spiritual growth and awakening. Thus I conceive of the "guide" role as a training role. For, in truth, the guide and the dying person come together to use one another for their own work on themselves. It is a truly collaborative dance between two people. The abstract point of this is that you don't do anything for anybody else, anyway. Actually, people do things to themselves and you are merely the environment in which they do it when they are ready. Thus the "guides" must be at a certain stage of their own evolution in order to be environments for awakening through death. They must have a connection with planes of consciousness beyond time and space that lead them to have a philosophical foundation that allows them to be equanimous and without panic in the face of death. They must view death not as an endpoint, but as a process of transformation.

At the level where there is only one of us it can get scary because that's where you are confronted with the cessation of yourself as a separate entity. It is spiritual practices such as meditation which help you to slowly extricate yourself from attachment to the levels of illusion of your separateness. Until you are free of that illusion and can merge into the ocean of existence without fear, everything you do subtly perpetuates the illusion of separateness. As long as you are attached to your separateness, you can't help but perpetuate fear, because there is a subtle fear in you of losing your separate identity. Because most of us have not fully realized our unity with the cosmos, we must continue to work on ourselves, and also continue to serve our fellow sentient beings to relieve suffering. Yet how can we remove suffering when we ourselves still fear? It is a matter of degree. We do what we can for others and yet never stop working on ourselves, for we keep in mind that it is our consciousness that may help to liberate another.

In discussing a Center for Dying, I referred to those who seek to die consciously. There are a number of new experiments in providing dignity to the process of dying for the general population, most notably Hospice in Connecticut, modeled on work in England. These programs are excellent and have my blessings. However, I am at this time interested in a far more specific undertaking. The spiritual community in America is made up of individuals who seek to awaken spiritually, or consciously, through the experiences available in this lifetime. They are proportionately few in number, because most people are preoccupied either with survival or optimizing pleasure. This small proportion is represented in the population of people in the process of dying as well. They are beings who, because of their state of evolution, wish to use their death to help themselves awaken into the spirit. It is

117

these people whom I wish to contact and for whom I would create a Center for Dying, or, perhaps, a Center for Learning through Dying.

In time, were this model useful, it would become visible in the society. Such an emergent institution would through the power of its statement draw to itself another segment of the population who would only be awakened to the possibility of learning through dying by the existence of the institution.

About a year ago, as I was about to seek funding for such a Center, Stewart Brand of the *Whole Earth Catalog* and *Co-evolution Quarterly*, who is also interested in this field, asked, "Why do you need a center? Why don't you just start with a telephone?" Sort of "Dial-a-Death!" This would allow people to call and ask for help in dealing with their own death. Guides and other assistance would be provided them in their own homes or at hospitals or wherever. We want to bring death out of the hospital environment, as has happened with birth in the last ten years. Or at least, help to create hospital environments that enable you to die consciously.

As we bring death out from under wraps as we are doing with birth, we shall be a stronger people for our ability to live with the truth of nature. But *compassionate* use of truth requires discretion. In dealing with death you must be prepared to speak the truth when someone asks you for truth. By the same token, you must remain silent when somebody is trying to deny their impending death. My father is seventy-six years old and he spent a great deal of time worrying about death, because he had gone beyond the actuarial chart point, although he was still quite healthy. I would give a seminar about dying out on the lawn of his home with a couple of hundred people in attendance. He'd come out and listen for a few minutes and then go in and watch the ball game. Then, when I went in later, he'd say, "Had a big crowd today." He was just not going to hear it. I would have liked to be able to say to him, "Let me share with you the thing that will relieve your anxiety, because you're my father and I love you." But, in another sense, we were just two beings who happened to be in an incarnation this time around where he was my father and I was his son and our approaches to life were very different. I used to come on to him, then I stopped. My compassion had ripened a bit and I could just love him as he was—in his perfection.

More recently he has asked several times about death and meditation. And a few weeks ago at dinner he said, "Because of our conversations this year, I don't seem to be as concerned about death as I used to be. I find at times I'm almost looking forward to it."

There is a book entitled *Life After Life* which Elisabeth Kubler-Ross introduced me to. The book concerns one hundred and fifty cases of individuals who were pronounced clinically dead and then were revived. Some at accident scenes, some in hospital operating rooms. Many of them reported experiences during the time they were supposedly dead of floating, reviewing their lives, meeting friends or relatives who had previously died, and meeting a being of light. The power of this data is in the similarities of their reported experiences. Studies of this type are a significant step in bringing life after death into the purview of the Western mind and thus dissipating anxiety about death.

My visits with Dr. Ross have greatly enriched my appreciation of the possibilities of working with dying people. Though a highly competent general practitioner and psychiatrist, this diminutive Swiss lady has had many profound experiences with what lies beyond death, and conveys a deep peace and reassurance to her patients.

Part of being conscious is in not trying to impose a limited rational model on how the world is, but rather in realizing that the rational model is a finite sub-system and that the law of the universe is infinite. To work with someone who is dying in the face of these paradoxes is to see the perfection of the dying and at the same moment to work full-time to relieve the suffering involved and to prepare the person for the moment of death.

In the Eastern tradition, the state of your con-

119

sciousness at the last moment of life is so crucial that you spend your whole life preparing for the moment. We've had many assassinations in our culture recently and when we think what it was like for Bobby Kennedy or Jack Kennedy, if they had any thought, what it would have been. "Oh, I've been shot!" or "He did it," or "Goodbye," or "Get him," or "Forgive him." Mahatma Gandhi walked out into a garden to give a press conference and a gunman shot him three or four times, but as he was falling the only thing that came out of his mouth was, "Ram. . . ." The name of God. He was ready!

At the moment of death you let go lightly, you go out into the light, towards the One, towards God. The only thing that died, after all, was another set of thoughts of who you were this time around.

Freeing the Mind

Now you're beginning to get a sense of the totality of the sadhana, the practice. We are working with the heart to open a flow with forms in the universe including thoughts and emotions. And that oceanic flow ultimately takes form and converts it back into energy. And that flow is also used to offer up stuff to get rid of it.

The offering up or the consuming or the cleaning is called purification. It exists in every religion. In Raja Yoga that's called yama. The various vows you take in Buddhism, the abstinences and commandments in Christianity and Judaism. These are done out of what is called "discriminative awareness." That is, you understand that you are an entity passing through a life in which the entire life drama is a feeding for your awakening. You see that the life experience is a vehicle for coming to God, or becoming conscious, for becoming liberated. And you understand ultimately that's what you're doing here.

When you're really wanting God, not just wanting to want God, you understand that is all you are doing here. When you are only wanting to want God, you say, "Well, I have this and that to do but I'd like to live my life in order to return to God." If you have studied at any depth the four truths of Buddha, you understand that the liberation from clinging and attachment is the liberation from suffering, and that liberates all beings from suffering. Or if you are thinking, "Well, I can't just want to go to God, I must help

121

other human beings as well," you begin to see that these are not polarized, that these are very intimately integrated. Because every act you perform for another human being can liberate them to the extent that you are liberated. If you feed someone with attachment, you fill their belly but you also reinforce their attachments. And that reinforcement perpetuates their long-term suffering. And thus ultimately when you understand that every act you do in life becomes an act of work on yourself because that is the highest thing you can do for all sentient beings; whether you're feeding somebody, or sitting in a cave meditating, or making people laugh, or providing a service or goods, or making a sandal. Whatever you're doing, you're a transmission.

For instance the image we've had of musicians is that of entertainers: you are playing music for other people to hear. As you begin to more clearly hear what the dance is about, you understand you are a soul on the journey towards your own liberation, and everything is grist for the mill, including your flute and your flute-playing. Then you are doing what the *Bhagavad-Gita* talks about; you are playing the flute as an offering to God. So you're playing it back into the circle. And as you use the flute-playing as an act of purification, it's no longer 'my' flute and it is not played for ego gratification. It's part of sadhana, it's an offering; the whole process is an offering. Then the flute playing starts to get pure and come from a higher and higher space. If you listen to, for example, Bishmella Khan play the shanai, you will hear a being who is playing to God, and that circle is so closed that it's God playing to Itself. Other human beings simply listen in. And because it's played that way, the subtle consciousness of the artist doesn't suck another person into "being entertained." Because the separation between entertainer and entertained is a distinction between subject and object, it's a distance between human beings. When you are part of the instrument of the music of God just playing to Itself, then anyone can tune in and become part of that same circle; and there's only one of it. There's no separation. The minute you think you're entertaining somebody, the minute you think you're feeding somebody, the minute you think there's a "them" out there, you just lost it. You just stopped the flow. Your concepts stop the flow. Because the mind, the thinking mind, works in relation to subject and object. And the minute you think about who somebody else is, the minute you define that person as somebody who needs food, and that's the reality, the only reality, you have made them into an object. No matter how much you feed them you've still got them separate from you. Most people who give in America, give out of paranoia and fear; it's like giving to keep people away.

It's very far out because I remember lecturing, I believe in Portland in the civic community center though usually I lecture in more funky spaces, and there was a big orchestra pit and the audience was "out there" and you couldn't even get the lights up high enough to see them. The whole game was designed to make everybody object and subject. There was a great barrier like a moat to protect you from the masses. It was incredible! And the insurance laws didn't allow me to have anybody on stage. I was alone on the stage with Krishna Das—a stage large enough to hold a whole opera company. And there's a sea of "them" out there who have come to be entertained; which is the traditional way of show business. But the cutting through of that doesn't necessarily demand a change in the physical game. It would have been nice if everybody was sitting around me, close and friendly. But it isn't critical. It's where my head is at. And it's where your head is at in every act you perform that determines whether that act liberates or entraps you and everybody around you.

Now we're approaching the issue of no-mind. How do you use your thoughts and how do you transcend them? And how do you go beyond mind? When you acknowledge that your life is a vehicle for your liberation it becomes clear that all of your life experiences are the optimum experience you need in order to

123

awaken. And the minute you perceive them that way they are useful within that domain. The minute you ignore that perception, they won't work that way.

In psychology there's a term called "functional fixedness." It's like you look at a hammer and you think of a hammer as an instrument to hammer nails. You might need something to serve as a pendulum, but every time you look at the hammer you only see something to hammer nails. The idea that the hammer could be attached to a string and used as a pendulum doesn't come through because you can't break the functional fixedness; you've got a set about how it's supposed to be and you can't break that set.

Well, it's the same thing with life experiences. A culture has a set about what the meaning of experiences are all about. Like what death is about or what deviant behavior is about. Is it insanity, or is it mystical wisdom? There are all these models in a culture, all its functional fixedness that you absorb as to what your life experiences are about. Is it gratifying, is it pleasant or painful? The model this culture works within is a model of gratification through external agents, getting more from the environment, man over nature, control and mastery for gratification, for creating your own personal heaven in which your ego stays paramount, your ego is "God." And that's different from a culture such as the Hopi Indians where you hear about a balance of man and nature, harmony, the Tao, the flow. Man *in* nature rather than man *over* nature. It's not mastery and control; it's listening to hear the way in which you play a part in flow. Then it's not just your personal gratification; it's you being part of a process that transcends your own separateness. That's a way of talking about God, that's God in form.

So the models you have of your experiences in your mind determine whether that experience can liberate you or will continue to entrap you. This is the beginning of the use of the mind. This is discriminative awareness. That is, making the discrimination between those things, or those ways of looking at things, that

124

will bring you close to God, to your own freedom; and those things which take you away from it.

Now at the stage that many people I meet are at they do their practice, their method, as "good" and as well as they can. And then they take a little time off. They say, "Well that's been great; now what do you say we have a pizza and a beer and listen to some good music." Now that—pizza, beer and music—could do it for them too except in their mind there's a model that the "time off" has nothing to do with it. We've got these models in our head about what's going to get us there. Meditation we suppose will get us there, pizza we presume will not. But the pizza, beer and rock music could do it for you if you were open to the flow of it, but the meditation might not if you're busy being righteous about it. Because there's no act in and of itself that is either dharmic or adharmic. It's who's doing it and why they're doing it that determines whether it's wholesome or not. To stick a knife in somebody can be adharmic, taking people away from God, or if you're a surgeon it could be bringing them to God. Obviously the act of the knife into the person isn't the issue. It's who's doing what, how. Even two surgeons could use that knife differently, one dharmically, the other adharmically; yet both think they're trying to save somebody. One is tuned to God's will and one is not. One is on an ego trip of "I'll save you," and one is healing in the way of things, and says, "If it be Thy will, O Lord."

The statement, "If it be Thy will, O Lord," is "If it be in the nature of things, if it be in the Natural Law, if it be in the flow of perfection of form." God, or the Natural Law, or the Divine Law, does not have to be judged by you. It has to be understood by you and heard by you and felt by you, and be'd by you.

So you ask yourself, "How do I use my every moment to get there?" Not heavy, tight, "I've got to be careful, I might make a mistake." Light, dancing, trusting, quieting, flowing. It's got to be done with the flow of love and the quietness of mind. It's like the women in India who go to the well and come back with a jug full of water on their heads. They're talking and gossiping as they walk, but they never forget the jug of water on their head. The jug of water is what your journey is all about. In the course of it you do what you do in life, but you don't forget the jug of water. You don't forget what it's all about. You keep your eye on the mark. At first you have to prime the pump a little bit to do it; and you keep forgetting and remembering and forgetting and remembering. That's what the illusion is. The illusion keeps pulling you back into forgetting. Lost in your melodrama: my love life, my child, my livelihood, my gratification. "Somebody ripped off my stereo," "I don't have a thing to wear," "Am I getting enough sex?" . . . just more and more stuff. And you keep forgetting into it.

And then every now and then you remember. You sit down and meditate or you read Ramakrishna or Ramana Maharshi and suddenly, "Oh, yeah right; whew! That's what it was about." And you remember again. And then a moment later you forget. But what happens is the balance shifts. If you can imagine a wheel whose rim is the cycle of births and deaths, all of the "stuff" of life, conditioned reality, and whose center is perfect flow, formless no-mind, the source. You've got one foot with most of your weight on the circumference of the wheel, and one foot tentatively on the center. That's the beginning of awakening. And you come in and you sit down and meditate and suddenly there's a moment when you feel the perfection of your being and your connection. And even beyond that, you just are. You're just like a tree or a stream. There's only a second or two of it there at the hub. Then your weight goes back on the outside of the wheel. Over and over and over this happens. Slowly, slowly the weight shifts. Then the weight shifts just enough so that there is a slight predominance on the center of the wheel, and you find that you naturally just want to sit down and be quiet. That you don't have to say, "I've got to meditate now." Or, "I've got to read a holy book." Or, "I've got to turn off the television set." Or, I've got to do anything. It doesn't become that kind of a discipline any more. The balance has shifted. And you keep allowing your life to become more and more simple. More and more harmonious. And less and less are you grabbing at this and pushing that away.

You are listening to hear how it is rather than imposing a structure because you see that if you keep imposing structures it doesn't get you freer. And you begin to forget your own romantic story-line. "Who am I becoming?" "What will I be when I grow up?" All of these models just fall away. You just start to sit simply, live simply, be where you are, be with who you're with when you're with them; you hear your dharma. If it's making shoes, you make shoes. And you're making the shoe with your consciousness in the present, simple and easy, not having a fantasy of surfing in Hawaii. You're just making a shoe. And because of the consciousness that you're bringing into that making of a shoe, the no-mind quality of it, it becomes the perfect statement of *shoe-ness* that can come through you in view of your skills at that moment.

When you can just make a shoe while you make a shoe you *are* the meditation. There's nothing to do. Your whole life is a meditative act. There's no time you leave meditation. It's not just sitting on your meditation pillow, your zafu. All of life is a big zafu; no matter whether you're driving or making love or whatever you're doing. It's all meditation. It becomes interesting to reflect on your life as to which acts can be done from the zafu and which can't. Which acts would fall away were they done from this space of just clear, quiet presence? It's just a natural shedding that occurs as part of the process which we're all in.

Which brings us to an interesting space dealing with free will and determinism. In truth, I can not yet hear this issue fully and clearly. But I will share with you what I thus far can hear. Within the perfection of this divine plan is included the freedom of an individual to choose to be harmonious with, or to go against, the law. The way that was depicted in the Bible was Adam and Eve's choice to eat the apple. God, we can say, is the word which describes, which symbolically represents, that Divine Law that says, "Live here in the perfection of the flow, but refrain from eating the apple." But the choice whether you want to eat the apple or not exists within the Garden of Eden as well as everywhere else. The apple represents the separation of the individual from the flow in his own mind. The subject-object, self-conscious reality. That is knowing it rather than being it. Chomp! The eating of the apple. Separation.

As was said before, after the separation God looks at Adam and Eve and they're wearing fig leaves over their genitals and he says, "Who told you you were naked?" Because if you were one with the flow, why would you have shame or separateness or any of those things? You've all experienced the innocence in which nakedness is just pure flow and beauty. And you have all experienced shame. And you've been with babies and you've seen the freedom in that flow. And you all have yearned to have that flow back again. Being the flow is that innocence.

Though all along the choice is available, until the lifetime in which you begin to awaken, you don't identify with other than that which is totally determined. Until then, you identify with your thoughts, but your thoughts are all lawfully determined. They are all within the laws of cause and effect. But who you really are is not your thoughts. The you that you thought you were turns out to be a conditioned, mechanical process of body and thought and the "you" who you really are goes back into the Void, the flow, the Dharma itself. So the optimum strategy is to act as if you have free choice. And choose always that which you feel is most in harmony with the way of things.

One way of saying it is that before you awaken you are determined. It is totally man's will. Once you awaken, you are free to choose between man's will and God's will. Free to choose to look up or not to look up. That is one way of saying it.

For example, you think that you picked up this book out of some kind of free choice. But your interests and economics and intellectual capabilities are all products of certain previous conditions. In fact there are beings with the ability to get outside of time who can see that this is the choice you would have made because they can see the way in which the laws work and out of what "stuff" that choice came. In that sense, that was not totally free choice. And yet it's not fatalistic.

The issues of determinism and free will—fatalism, karma, dependent origination—all weave an incredibly complex pattern that I think it would be somewhat beyond the scope of this work to try to spell out. So I'm saying to you at the moment, act as if you are a free agent and choose to awaken. In other words, you have real discriminative awareness to use. A skillful use of the intellect is contemplation. Every morning work with a thought. Take a holy book. Don't read pages, don't collect it; take one thought and just sit with it for about ten or fifteen minutes.

You could contemplate on the qualities of Christ. Charity. Suffering. Every day you contemplate on the stuff you're becoming. Sri Ramakrishna said, "If you meditate on your ideal; you will acquire its nature. If you think of God day and night, you will acquire the nature of God."

Fill your mind with things that are going to get you there. Your mind doesn't have to be filled with the daily news to prove that you're a good citizen. You don't have to be at the mercy of all this, the constant onslaught of media. You could fill your mind instead with the stuff that liberates you. Ultimately becoming aware of that which gets you to God and that which doesn't, to help you let go of that which doesn't.

You begin to develop the power of your mind through concentration, through one-pointedness. Following the breath, following the mantra, whatever is your dharmic choice. You develop the capacity to put your mind on one thought and keep it there and let everything else flow by. You don't stop your mind. You let it flow. But you bring one thought constantly to the surface. You keep coming back to one thought all the time. Breathing in, breathing out. Breathing in, breathing out. Rising, falling. You note breathing in, breathing out; or you use your mantra, "Ram, Ram, Ram, Ram, Ram, Ram, Ram, Ram, Ram, Ram." Eating, sleeping, making love . . . "Ram, Ram." You "Ram-ize" it. You convert it all by maintaining a frame of reference. That has the dual capacity of centering you and increasing the power of one-pointedness. A one-pointed mind is free of the intellect, it is a supple, useful mind.

You see, you can use your intellect to judge the universe or to clean up your own game. If you judge the universe, you are using your intellect to take you away from God; if you use it to clean up your own game, it can take you towards God, towards the Tao, the way of things, the Divine Plan as discussed in Judaism, Christianity, the Moslem religion, Hinduism, Zoroastrianism; you could call it the Mind of God, you could call it the Natural Law, you could call it the way in which everything in form is related to everything else: that is the flow; and that flow is harmonious in its parts. Even the cacaphonous parts are harmonious, in the larger scope of things. They are not lawful in linear, analytic, logical sense. Natural Law includes paradox, which logical law cannot. "A" can be "A" and "not-A" at the same moment. It's not a law that you can grok with your intellect. It's a law that you can become, but you can't know. The closest we come to a sensing of the law is what we call intuitive wisdom in the West. Gurdjieff called it "the higher faculty." It's a higher way of knowing, a subjective involvement in the universe; not an objective one. You don't know the law; you are the law. And you sense, when you have a quiet mind, the way of things.

Just as some American Indian tribes would send a pubescent boy out into the wilderness for a few days or weeks to fast and listen, to become quiet, and attune to the way of things; so it is necessary to get quiet enough to hear, not only the singing mating calls of the birds, but the way of your own sexual desires, the way of your own patterns of anger, the way of your own heart, the way of the decay of your own body, without getting lost in grabbing hold, in judgment or analysis or clinging or fear; but just hearing it as it is.

It's not the objective "witness" in the sense of standing back and looking. It's a subjective being part of it without attachment anywhere. It's a very subtle place I'm talking about now. It's the use of the mind beyond the intellect. The intellect is the first step of it, discriminative awareness; looking around and saying, "This anger isn't going to get me to God, I'm going to drop it." You drop it because you see where you're going this lifetime. It's like going to New York City and you come to a road which leads to Mexico, but you don't take it this time. Mexico is beautiful but it's not where you're going this time around.

Discriminative awareness is based on goal-oriented behavior. But as you get near the end of the journey, you must give up even the concepts of the goal, and of the trying, and of you being someone seeking. Because even those concepts ultimately keep you back. All concepts, all models, all molds, all programs in your head, are limiting conditions. No-mind, the sufficient faith to exist in no-mind, to just be empty and trust that as a situation arises out of you will come what is necessary to deal with that situation including the use of your intellect where appropriate. Your intellect need not be constantly held on to to keep reassuring you that you know where you're at, out of fear of loss of control. Ultimately, when you stop identifying so much with your physical body and with your psychological entity, that anxiety starts to disintegrate. And you start to define yourself as in flow with the universe; and whatever comes along—death, life, joy, sadness— is grist for the mill of awakening. Not "this" versus "that," but "whatever."

Under those conditions you don't have to do so much labeling. You can just be quiet and let the universe happen. But the trust in that is based on the giving up of your own unworthiness. Because if you think you need your mind to keep you under control, because if you lost control you would become a wild, destructive, chaotic, uncaring, insensitive being; you are defining yourself as did Freud as totally selfish behavior. But the predicament is, that's defining your existence from just the first two chakras. Even when

131

Adler comes along and says the real guts of the human being is power tripping, that too is only the third chakra. But above all that is the heart which corresponds seeming opposites and brings it all into flowing understanding and acceptance. And yet above that you also exist in the fifth, sixth and seventh chakras as well.

And once you begin to awaken and sense who you are, you begin to understand how you are becoming the Dharma. Now as you attune you literally cannot screw another human being. You cannot go out and do them in because not only your intellect understands the karmic implication of it, but your perception is such that you see yourself screwing yourself. The concept of brotherhood is no longer an intellectual, liberal concept; its' a perceptual reality. And living within that reality it's impossible to perform certain acts you were performing before. And that's what the Ten Commandments are about. They are a statement of how it is when you see things as they are. But because most exoteric religions are written for people who are not awakened, they are used as moral prescriptions using guilt in order to control behavior, to move people slowly in that direction.

You must become in your own life the living statement of the Vedas, of the Commandments, of the Law. A conscious human being *is* the Law, *is* the Dharma. You don't know the Dharma, or recite the Dharma. You are the Dharma. Your every act. The way a Roshi washes a dish is the Dharma. That washing of the dish is in perfect harmony with all the forces in the universe at that moment. No mind involved, no analytic thought, "Am I doing the right thing?" In discriminative awareness, the intellect is used only in the early stages. Later, there is no-mind. The intellect is available as a servant, but not as a master. It is available to do analytic work when you need it. It's as beautiful and powerful an instrument as your prehensile skill, as your ability to oppose thumb and index finger, an ability you're delighted you have. If you didn't have that prehensile ability it would change your life considerably. But you don't have to go around all day picking up things just to keep showing you can do it. I mean the awe for it diminishes after awhile. It's a power, a siddhi, that you have because of your simian nature. Apes have it too. So, too, your cerebral cortex is a power, because of your homo sapien nature. You can sit around and flex it, in the *New York Times* or wherever you wish, to the delight of everybody. Fascinated with your own power. Going to the moon is a projection of our human intellect. Man over nature. Though we worship the human intellect, as an exquisite power, it is very trivial in the greater design of the natural law of things.

The question is: Are you going to play big league or do you want to play sandlot ball? That's really what it boils down to. In big league the intellect is available. I'm no more stupid than I ever was. My mind is perfectly good; as good as it was when I worshipped it as a professor at Harvard. But it's sure not a very big part of my life. And even at this moment as all this stuff is coming out, I'm enjoying it as much as you are, it's coming out of a place of total emptiness in me. I couldn't care less. It's coming out because this situation is eliciting it. Because our collective mind is eliciting this kind of clarification of our predicament at this moment. It's dharmically appropriate for this moment. I have no ego investment in this stuff; because it isn't mine. If you don't deviate the flow or color it with your own trip it comes through purely in whatever form it is your dharma to express, and the mind is freed.

⚜

Nobody Special

We are in training to be nobody special. And it is in that nobody-specialness that we can be anybody. The fatigue, the neurosis, the anxiety, the fear, all comes from identifying with the somebody-ness. But you have to start somewhere. It does seem that you have to be somebody before you can become nobody. If you started out being nobody at the beginning of this incarnation, you probably wouldn't have made it this far. Blue babies are examples of nobody special. They just don't have the will to breathe or eat or live. For it's that force of somebody-ness that develops the social and physical survival mechanisms. It's only now, having evolved to this point, that we learn to put that somebody-ness, that whole survival kit, which is called the ego, into perspective.

When I was a Harvard professor, I would spend all my time thinking. I was paid for that. I would have clipboards and tape recorders to collect all my thoughts. Now I've become very simple. My mind is quite empty. There is nothing in there at all, and I just sit looking stupid. Then when something needs to happen, it happens, and I don't have to listen to it.

It's very far out when you begin not to think, or the thinking is going by and you're not identified with being the thinker. At first you really "think" you've lost something. It's a while before you can appreciate the peace that comes from the simplicity of no-mind, of just emptiness, of not having to be somebody all the time. You've been somebody long enough. You spent the first half of your life becoming somebody. Now you can work on becoming nobody, which is really somebody.

For when you become nobody there is no tension, no pretense, no one trying to be anyone or anything, and the natural state of the mind shines through un-obstructed—the natural state of the mind is pure love which is not other than pure awareness. Can you imag-ine when you become that place you've only touched through your meditations? When you *are* love. You've finally acknowledged who you really are. You've cleared away all of the mind trips that kept you being who you thought you were. Now, everybody you look at you're in love with. You experience the exquisite-ness of being in love with everybody and not having to do anything about it. Because you've developed compassion. The compassion is to let people be as they need to be without coming on to them. The only time you come on to people is when their actions are limit-ing the opportunities for other human beings to be free. And then the way in which you come on is very mind-fully and open hearted. For if you are somebody coming on to change someone, you're just creating more anger. If you are nobody special but it is your dharma to come on about injustice, then it is merely an act of the Dharma. And not for a moment do you lose that total love for the other person who is not other than you. For being nobody, there is nobody you're not.

Had you sufficient discipline you could pursue the steepest of paths to get rid of all the ways you cling to models of yourself. You could just sit—Zen Buddhism —and every thought that comes by, that creates another reality, you would let it go. And clinging to none, you would know enlightenment. Or you might pursue the path of Ramana Maharshi, Vicharya Atman, "Who am I?" You simply ask "Who am I?" Who am I?" And slowly you watch yourself be other than all the ways in which you identify yourself—as bodies, or-gans, emotions, social roles—you see it all. You keep dissociating from it until you are left with the thought of "I." "I am the thought 'I.'" This path takes incredible discipline for as you have freed yourself from your body and your emotions and you're just about to drop this last thought of "I," your body grabs you again. And you're back in your habitual thoughts about your body, your identity.

Most of the time when you watch your mind you find it keeps grabbing at things and making them the foreground. And everything else becomes the background. When you're reading, you're not listening. When you're listening, you're not seeing. When you're remembering, you forget where you're at. But can you function when the world is all background and awareness itself is foreground?

When awareness is identified with thoughts you only exist in a certain time/space dimension. But when awareness goes behind thought, you are able to be free of time and see thoughts appearing and disappearing, just watching thought forms come into existence, exist, and pass away in a millisecond. And when the intensity of concentration allows you to see the space between two thoughts, you see eternity. There is no thought there. You realize that thoughts exist against the backdrop of no thought. Against the backdrop of emptiness, of nothing, we exist. And there you are at the edge of perceiving who you are. Then you face one of the greatest fears you will ever confront: the fear of your own extinction. The fear of ceasing to exist not just as a body, but even as a soul. It is similar to the statement made by Huang Po about people approaching this point that they become fearful to enter into what they consider "the void," distressed that once they let go into it they will drop unendingly, that there will be nothing to stay their fall, not realizing the Void is the Dharma itself.

But as you're ready for the ultimate mystic doorway, the inner door of the seventh temple, you say, "I am not this thought." You let go of even the great fear of non-existence. The senses are just working by themselves. There is hearing occurring but there is no listener, there is seeing but there is no see-er. The senses are just all doing their thing but there's nobody home. If the mind thinks, "I am aware," that is recognized as just another thought, a part of the show passing by, it's not awareness itself. Thoughts are going by like a river and awareness simply is. When you become just awareness, there is no more "you" being aware.

By letting go of even the thought "I," what is left? There is nowhere to stand and no one to stand there. No separation anywhere. Pure awareness. Neither this, nor that. Just clarity and being.

Karmuppance

In the mid-sixties there seemed to be an expectation that if you got high you'd be free. It was not quite realistic about the profundity of man's attachments and deep clingings. There was a feeling that if only we knew how to get high the right way, we wouldn't come down. And that was our attempt. Then in the late sixties there was the idea that if you joined the movement and became part of a model of how to stay high, you'd be able to do it. So in the late sixties and early seventies there was a tremendous interest in mass movements.

Now people are realizing that it's somewhat of a long haul. They're feeling transformations in themselves, but they're working with their lows as well as their highs, they're cleaning up their game. And the reason we clean stuff away and don't just get high, why we focus on our depression and our negativity and all of our heavies is because we're getting hip to the fact that, if we push stuff under the rug, sooner or later there is karmuppance.

For instance, recently I was invited to visit "death row" in San Quentin. To be as honest as I try always to be with you, and with me, I sat outside the prison before I went in, in my rented car, looking at San Quentin, thinking, "I'll be happy to go in; and I'll be happy to come out." Because there is a certain kind of paranoia in the searching procedures and the authority structure that I have to keep consuming. I went in and was met by all the yogis who teach there and the acting warden who was a very nice guy. And we were immediately whisked up to death row. There are two rows actually because there were so many of these fellows, they are in separate cells, segregated in two long rows separated by a wall.

These men are in a peculiar predicament. The death penalty has been reconsidered by the Federal Supreme Court. And the death penalty has been re-instituted. California has already voted to resume the death penalty, so if the State Supreme Court concurs some of these men will die within sixty to ninety days.

As I went up to each cell, out of the thirty-four men, there were not more than five who did not receive me openly, clearly, quietly, consciously. The feeling I had was that I was visiting a monastery and that these were monks in their cells. For these men, who are facing death, have been pushed into a situation that has cut through their melodrama and they are right here. We sat together in groups of ten and as part of the meditation we were sending out thought forms of love and peace to all sentient beings in the universe. I became so affected by the vibration of the space that it was very hard for me to move on to the next group. There was light pouring out of these beings' eyes.

And we got so open that I was able to say, without any of us freaking, "I can't tell whether what's happened to you is a blessing or a curse, for there is very little chance that we would be sharing this high a space or even would have met were you not in this situation." To prove my point, I'll tell you that I spent half an hour on one of the other segregated main-line cell-blocks. And of these beings the percentage of those open were just what you'd expect in our society. Maybe one out of a hundred was right there with me. From the rest you could feel the cynicism, the doubt, the put-down, the sarcasm. Now the bizarre humor of all this is that if the Supreme Court rulings stop the death penalty, these men will all become lifers and almost all of them will lose this consciousness. Yet if they die, they will have this consciousness right up to the moment they enter the gas chamber; which does

139

not mean that all the karma accrued to them—because in most cases they have been involved in killing another human being—is over.

Because you can go into death with "Ram" on your lips, with Christ in your heart, high and clear; but whatever stuff is covered over by your situational high at the time of dying, as your ego structure starts to lose its control, the stuff that's left will bubble up again and you are going to have your karmuppance, you will once again renew your karmic run-through.

There is a story about an old Zen monk who was dying, who had finished everything and was about to get off the wheel. He was just floating away, free and in his pure Buddha-mind when a thought passes by of a beautiful deer he had once seen in a field. And he held onto that thought for just a second because of its beauty and immediately took birth again as a deer. It's as subtle as that. You can't cheat the game by getting high is the point. The situation these fellows are in is forcing their openness and awareness, but it's not totally burning out their karma. It will help. One moment in which they feel compassion for the person they may have murdered will do much for their karma; but it's not going to purify all of it.

It's like when you begin to see the work that is to be done and you go to an ashram or a monastery or you hang out with satsang. You surround yourself with a community of beings who think the way you think. And then none of the stuff, the really hairy stuff inside yourself, comes up. It all gets pushed underground. You can sit in a temple or a cave in India and get so holy, so clear and radiant, the light is pouring out of you. But when you come out of that cave, when you leave that supportive structure which worked with your strengths but seldom confronted you with your weaknesses, your old habit patterns tend to reappear, and you come back into the same old games, the games you were sure you had finished with. Because there were uncooked seeds. Seeds of desires that sprout again the minute they are stimulated. You can stay in very holy places and the seeds sit there dormant and uncooked. But there is fear in such an individual because they know they're still vulnerable.

Nothing goes under the rug. You can't hide in your highness, any more than you've hidden in your unworthiness. If you have finally decided you want to go to God, you've got to give it all up. The process is one of keeping the ground as you go up, so you always have ground, so that you're high and low at the same moment—that's a tough game to learn, but it's a very important one. So at the same moment that if I could I would like to take you all up higher and higher, you see that the game isn't to get high, the game is to get balanced and liberated.

Most of us find that the veils of the illusion, of the clinging, are very thick and we want to do things to burn up these veils, to purify ourselves, and get on with it. And even though the whole model of getting on with it and going from here to there is itself a trap, we still can skillfully use that trap to clear away other obstacles that are hindering us, then ultimately we can trap the attachment to the method itself.

We are coming out of a cultural tradition where once you saw where you wanted to go, you took the most direct and aggressive path to get there. And impatience is part of the quality of our tradition. It's what made our country great. But the predicament you face is that the beginning of this awakening of who you aren't comes long before you are really ready to let go of all of the ways in which you cling, and some of these methods just become very powerful means of up-leveling old games, of reinforcing heavy ego trips. It's like, I know people who have meditated for years and they wear their methods like merit badges. "I've done six Vipassana courses, three seshins and a double dervish. I get up at four every morning. I can sit without moving for hours. My mind goes absolutely blank." They're professional meditators. They have to some degree mastered their method but they have not loosened the hold of grasping and greed. Their method has just become another form of worldliness. Nothing much is happening because it's such an ego trip. There are, for instance, people who can go into samadhi and stay there for long periods but when they come back they're no wiser than when they entered that state. It's like the story of the king who promised a yogi the best horse in the kingdom if he could go into deep samadhi and be buried alive for a year. So they buried the yogi, but in the course of the year the kingdom was overthrown and nobody remembered to dig up the yogi. About ten years later someone came across the yogi still in his deep trance and whispered "Om" in his ear and he was roused. And the first thing he said was, "Where's my horse?"

Spiritual work can be like gambling on a game of roulette. You put your money down and the ball goes around and drops into the slot your money was on. And they say, "Do you want to take your money or let it ride?" Anywhere on this journey you can take your money and pull out and go spend it. Or you can let it ride. Do you want to just double your money or do you want to go for broke? Do you just want a little social leverage or do you want to get done? It's no different than Mara confronting Buddha as he sits under the bodhi tree. For as you get closer to the inner gates of freedom, of enlightenment, of liberation, the subtle clingings will be fanned all the more and the opportunities for gratification keep increasing. Because of the one-pointedness developed through meditation, you become able to cut through your own limits of consciousness and see some of what it's all about. But if you have power needs, you are then all too ready to use what you see to have power over other beings. If your spiritual work has come out of wisdom, not out of a need for power, but out of a yearning for God, then when the powers come you just notice them, realizing they are going to take you on tangents, consume them and keep going. You just have to trust the light and let your money ride. For as long as you think you are "somebody," you aren't yet quiet enough to be in tune with all of it and thus any action taken is done from your own particular separate perspective.

As long as you are in an incarnation, there will be action. As long as there is form there will be change.

But it depends on who is doing the acting or thinks acting is being done that will determ whether that act is part of the flow of things antagonistic to it. It's like the story about the prince's butcher. The prince asked the butcher how, although he had been cutting with the same knife for nineteen years, it never needed to be sharpened. And the butcher explained that he is in tune with what he is cutting, that the knife finds its way into the joint, above the bone, through the muscle. That it doesn't hit against the joint, it just finds its way around the bone. Because he is tuned, he is what he is doing, he isn't busy being a butcher cutting a piece of meat— he is awareness, and that awareness includes the meat and the butcher and the knife. There is an act happening, but there is no doer of the act because there's nobody who thinks he's a butcher.

When you are in harmony with the way everything proceeds from everything else, you cannot act wrongly. For not only are you in tune with the particular act you are doing in terms of time, but in all of the ways in which that act is interrelated with everything in the universe. It is a level of awareness from which actions are manifest that have no clinging. Not even clinging to the effects of the act. You are not holding on anywhere. You're right here, always in the new existential moment. Moment to moment it's a new mind. No personal history. You just keep giving up your story line.

Each person gets their karmuppance. If you focus on God, you get God. If you want power, you get power. If you want more of something, you get it. The horror is that you get everything you want. And often when you finally get it, you don't want it. The process of karmic fruition speeds up because as you get closer you see yourself living out old karma, old desires. As your life gets freer and freer of attachments, you create less and less karma. For karma, or that which hangs onto you, is created by an act done with attachment. When you're not clinging to senses or to thoughts, you are not creating more karma, there is no one intending anything to happen in any way, there

; no one separate to act in a separate way. When there is no attachment or identification with thoughts and feelings, there is no reactive push into action creating more doing, more karma. Not identifying, not being separate, cooks these seeds and consumes the grasping for more.

You get to the point where your acts are not done out of attachment, but instead are just done as they're done and no new stuff is being created. There is just old stuff running off, but nobody being affected by it because there is nothing in you which clings to a model of who you are or aren't. It all becomes just passing show, there is no investment in it representing you as an "individual," it is just the outcome of previous input, just old conditioning clicking along, just more grist for the mill.

Recently, while I was with Elisabeth Kubler-Ross, she was invited to speak before a thousand neurosurgeons about dying. She described exquisitely the stages people go through as they approach death. The resistance, the denial, the bargaining, the anger, the despair, and then the opening space. That same opening which is on San Quentin's death row. Then she went on to describe what happens after death with data designed to convince the scientific community. And I said to her, "You have a hard row to hoe to convince those scientists. I decided a long time ago to just become it; not to prove anything to anybody." And she said, "Well, you and I have different dharma." She does it with no attachment, no clinging, very light, very dancing. And when I said to her, "Isn't it remarkable that I've lectured to over seventy thousand people this year and how open these beings are, Elisabeth, to be sharing this kind of consciousness." And she said, "Well, don't you understand we're all on death row."

God and Beyond

All the time I was with Maharaj-ji, he never had me
meditate. He'd feed me, love me, pat me, yell at me,
cajole me, bore me, fascinate me, perplex me, send
me away, draw me to him. Yet, when I told him I was
going to do a Buddhist meditation retreat he said to
me, "Bring your mind to one-pointedness and you will
know God." When somebody showed him a book in
which there was a picture of Kalu Rinpoche on one
side and him on the other he pointed to one picture
and said, "Buddha," and then pointed to the other and
said, "Buddha."

One morning when I was in Allahabad at a house
where Maharaj-ji was staying maybe fifteen or twenty
Indian devotees came to see him. About thirty Westerners sat around the outside of the circle. One of the
Indians that came in was obviously a very important
man. I never could get clear whether he was a Supreme Court judge or an administrative director of the
court. When he came in, I was very content being in
the back with all of the Westerners and watching the
whole process. Suddenly Maharaj-ji started to build up
my image to this man, saying, "This is Dr. Alpert from
America, he is a professor at Harvard . . . a great saint."
And the Supreme Court judge turned and said, "Well,
perhaps you'd like to visit the court." Now, I come
from a family of lawyers and I've spent more than
enough time in courts. I was in India to be with
Maharaj-ji and didn't want to visit the court, but I was
caught in my social propriety so I said, "Well, that

145

would be lovely." And then he said, "Well, tomorrow at ten?" And I felt I was being trapped from the abstract to the concrete. So I said, "Well, you'll have to ask my Guru," figuring that he would get me off the hook. But Maharaj-ji said, "If Ram Dass said it'd be lovely, it'll be lovely. He'll go at ten." And then he pointed at me like, "Watch it baby; you lie, you'll pay. Captain Karma will get you." So I went to the court and watched a murder trial and then went to the law library and the librarian was a great student of the *Ramayana*; you've got it all. But only when you start to acknowledge it is a bar review room where all the lawyers hang out. And all the lawyers saw me, a Westerner, being escorted by this very important man whom they were all being very obsequious to. So they came over and tried to discuss Nixon's China policy with me. At that moment it was of great concern to India, I had just read *Time* magazine so I was in a perfect position to be an expert. So I discussed power blocks, Russia, and alignments; I did a perfect snow job. When I came back, Maharaj-ji kept asking me, "Well, what happened at the court?" And every time I'd go to tell him, he'd tell me because he obviously had watched the whole process from some other level. So I thought it was over and I had learned my lesson. Well that evening the head of the law association came to have darshan with Maharaj-ji. And he said to me, "We were thinking that you might perhaps address the Rotary Club and the Honorary Legal Society." I thought, "Oh no, I'm going to end up on the creamed vegetable circuit." So I said, "I really don't want to, you'll have to ask Maharaj-ji." I didn't even get into being nice. I thought I'd be really truthful. But if he tells me I've got to do it, I'll do it. So he goes up and says, "Maharaj-ji, we would love to have Dr. Alpert address the Honorary Legal Society and the Rotary Club." Maharaj-ji looked delighted. When you knew Maharaj-ji, who sat with a blanket and a watering pot, and couldn't care less, you knew it was all nonsense from where he was sitting. But, oh, he was fascinated. And he was saying to everybody, "Ram Dass is going to speak at the Rotary Club," as if "This is it, we've

146

finally broken through, we're going big time now." And my heart was sinking. For half a moment I thought, "He just was hustling me, he wants to exploit me to become big time in India. Oh damn it, I've been had again."

And then he said to me, "Well what are you going to talk about?" He was terribly interested. And I said, "Well, I don't know Maharaj-ji. I guess I'll talk about Law as Dharma." I was grasping at something quick to be cute about. And he says, "Uh huh; are you going to talk about Hanuman?" And I said, "Oh, of course Maharaj-ji." He said, "That's good." And I saw the lawyer's face take on a peculiar change. And then Maharaj-ji said, "Are you going to talk about me?" "Of course Maharaj-ji," I said, "you're my Guru." "Well, that's good. Are you going to talk about Christ?" "Absolutely." So the lawyer said, "Well, we kind of thought he'd talk about Nixon's China policy." And Maharaj-ji turned to him and said, "Oh no, Ram Dass is not to be trusted about worldly things. He can only talk about God; that's all he's capable of talking about. Ram Dass only talks about God." I said, "That's right, I only talk about God." And the lawyer said, "Well, in that case, perhaps he shouldn't speak to the group. Maybe I'll have a few lawyers who are interested come by my house." Suddenly the whole thing lost its interest to a very worldly group of people.

And I thought, "Far out, I've just been given my instructions. All I've got to do is talk about God for the rest of my life and I'm protected. I don't have to get lost in all the worldly stuff." But I came back to the West and it's funny to talk about God in the West. It's not a hip concept. It's not that God is dead; it's just that God is not a viable concept.

I've been talking about God a lot these last years but it's tricky because I am more than superficially trained in Buddhism. And Buddhist philosophy does not really involve itself with the concept of God at any great length. I am very attracted to the simplicity and cleanness of Theravada and Zen Buddhism. So, on the one hand, I'm faced with the Zen part of myself which finds the concept of God an unnecessary addition to a simple universe. And on the other, my Guru who says, "Speak only about God."

Now the world, the universe, looks different as your consciousness shifts. Where most of us start from is that we are psychological beings identified with our psychological accumulations. We are emotional, thinking, feeling entities. In Buddhism you learn about anicca, dukkha, anatta, the changeability, unsatisfactoriness and emptiness of all phenomena. You learn about the impermanence of things, of thoughts. The passing nature of all states of being, feelings, concepts. You learn about the suffering that is caused by clinging to these concepts. And finally with anatta you find that even the concept of self must go. Whether that self is physical self, or psychological self, or astral self, or soul self, or the Eternal Self. Concepts are concepts; and concepts must go. And even the concept of enlightenment or Nirvana or of that which is beyond self is just another concept. So why would we as psychological beings who are here with all of our problems and melodramas and attempts and strivings and awakenings and all that—why would we want to buy into more concepts when the game is to get rid of them and go beyond concepts? God is a concept. Soul is a concept. And when I say you are a spiritual entity who has taken birth in order to work out your karma, there is no "you" in truth whose karma must be worked out. There is only an apparent grouping of events, one of which is the concept you have of yourself. And as it all disintegrates through deeper and deeper meditation, and more and more emptying, you disappear along with the universe into the Void.

Well, what I have come to understand is that my path involves my heart, involves flow. It can't come after the fact. It has to be the leading edge of my method. And in a devotional path, you work with forms in order to transform your own identity. And in the process you break the habits you've held as your realities and your own self-definition. And the new realities, the new concepts, you take on, because they were taken on intentionally, don't have the same hold over you that the old ones had. It's using a skillful means to get rid of one thing when later you will get rid of that aid as well. Ramana Maharshi refers to these concepts, specifically the concept of "I" or "Self," as the stick that you use to stir the funeral pyre. If you go to Benares you'll find the bodies being burned at the burning ghat and men with big sticks stirring the embers to make sure the whole thing gets burned. And after they finish stirring a particular fire, as it's getting near the end they throw the stick on the fire and it too burns up.

And so it is with Gurus, teachers, methods and God. For what God is, is beyond the concept of God. It's exactly the same thing as the "Gate Gate Paragate Parasamgate Bodhi Svaha" mantrum. It's beyond the concept of beyond. Now in Judaism, except in the Hasid tradition a bit, the closest you get to God is coming into His presence, but dualism remains right to the end. It's ultimately still dualistic, I–Thou. It is blasphemy within traditional Judaism to conceive of merging with God; for God is unknowable, it's G–d, it's unspeakable. It's the word that can't be spoken.

From my point of view as a heart being, as a devotional being, I have a Guru and there is Hanuman, there is Kali, there is Durga, there is Krishna, there is Ram, there is Jesus, there is Buddha, there is Rama-krishna, there is Ramana Maharshi. My universe is peopled with these beings. They are no realer than you are. The only difference between them and you is that they know they are not real and are free. You are still thinking, attached to what you think you think. And they aren't. They are what's called the sangha in Buddhism or satsang in Hinduism, the community of beings that I hang out with. There is nothing that comes out of them that is going to entrap me because they know it's all lila, all just the cosmic dance of being.

I sat before Maharaj-ji—before this man in a blanket—just loving him. All I wanted to do was caress his foot. It's extraordinary to love somebody that much. I was in ecstasy just looking at him. I'd been doing that for months and months and finally I thought, "Maybe this is just the veil; I've got to go beyond the veil." One day I was sitting across the courtyard from him and everybody was up around him and I thought, "I don't have to be around him. This is just form. Look they're all worshipping the form. The form isn't it. I don't have to be here." Just then Maharaj-ji sent an old man over to touch my feet. I asked him why he did it and he replied, "Maharaj-ji said go over and touch Ram Dass' feet because he and I understand each other perfectly." Because at that moment I was seeing through the method that I was using. I wasn't stopping my love for him, but I wasn't trapped by it anymore.

Now I could talk about the Dharmic Law, but it's very hard for many people to fall in love with a Dharmic Law. But it's very easy for me to fall in love with God. But the concept of God is very much the same as the concept of the Dharmic Law. It's the law of the relationship of things. It's the original consciousness; it's the one mind; it's the Ancient One. It's that which at its heart is empty. I'd sit before Maharaj-ji and think, "I'm not going to be taken by the form." So while he'd be handing out stuff I'd be meditating with my eyes closed, focused on my third eye and I would start to feel this change coming over my body and I'd feel more and more energy; and suddenly with my eyes closed I would meet him on a different astral plane. Now you can get fascinated with that. That's the whole world of the occult; all those forces and beings to play with. "But Maharaj-ji, that isn't who you are either." And I'd go right through that one. It's like going through infinite doorways. You come to another one and you think that door is the final temple; and then you see it's just another doorway and you go through it. And you go through another and another. And if you go through far enough you come right back to yourself, which isn't either. The whole thing, method and all, just disintegrates before your eyes.

When I sit with Maharaj-ji, my heart flows. I flow into the universe of forms, and the universe of forms flow into me. As that flow gets greater and greater, the boundaries between Maharaj-ji and me disappear. For

150

me, Maharaj-ji is the universe; so that the differentiation between me and the universe disappears into a flow of energy. As I open more and more through my heart and thus become more and more of the flow of the universe in its energy form, I start to rise. It's as if it's a fuel. And I rise into states of consciousness which are known as jhanic states or samadhi states. Each experience is another form of Maharaj-ji, or the Mother. And must be consumed. For Maharaj-ji must be consumed by me, taken into myself; I must surrender into him until there is no boundary. He is not only all the forms that are available to my physical eye, but to my spiritual eye as well. And as I go into higher and higher states, there are fewer and fewer forms. And many of the forms are only half-formed; for they are on the edge of where the form and formless meet. On the way through these planes are experiences of emptiness and coldness and impersonality. They are not empty because they are "experiences" of emptiness. That's different. There are planes or states of incredible bliss and rapture, where your whole body is writhing in delight. It's as though every cell were having an orgasm. There are states or planes of consciousness of diamond clarity, in which you see and know and understand everything's relationship to everything else. It's as if you are privy to the secrets of God. There are planes or states of consciousness where everything is aesthetically so perfect even the words come out as poetry and all is luminous and colorful. Aldous Huxley writes about that. Aldous once said, "The reason we like precious jewels so much is they remind us of planes of consciousness we've lived on where those are the pebbles." Form after form, plane after plane, state after state, experience after experience—all within Maharaj-ji. All within my love and my flow towards other. And there comes a point where the flow is so open and the boundaries so far erased that you and Maharaj-ji, or you and the Mother, or you and the form, become one, at every level of form, all the way up to pure, undifferentiated energy.

Were that the end, it would be so easy for our

minds to grasp; so easy for science to control. But all of that is just a doorway. All of that merely brings you to the edge of the lake. It is at the edge of the lake that you experience the presence of what lies beyond form. Yet there is no "what" that lies beyond form. For there is no beyond. For what *it* is contains all that is. You, at that moment, sit at the edge where you *are* the paradox. All of the forms disappear into the lake of emptiness and yet they are not lost. It's at the edge of the lake that someone whose path is the path of the heart will say, "I am experiencing the presence of God." For one more step into the lake and the experiencer and the experience have merged and you have become God; and the concept of God is long gone. As you merge into God, you have entered into what the Buddhists call Nirvana. The game is not to know God; the game is to be God. To be God is to be nobody; and yet there is nothing that you are not.

If you come back into form from having merged with God, you are in the world though not of it. You play the cosmic sport. You fill the forms though there is no one home, it is just more lila, the dance of God.

There are beings who have roamed, and do roam, this earth and other planes of consciousness who have entered into that ocean and returned. Their existence liberates all who recognize them. You may have in your midst such a being, but you would never know it because you are attached to the form. You might be like someone catching an apple from Maharaj-ji and failing to recognize that there is nobody throwing the apple. So while he is God beckoning from beyond, you get lost along the path. And because God does not exist in time, he doesn't push you. For, sooner or later, one lifetime or another, you get home. And when you get home, you will realize that you have always been God. That you created your own separateness for the sport. The difference between you and a Guru or a perfected being is that they aren't and you still think you are.

Christ said, "I have come to bring you to the Father. I am in the Father; the Father is in me. You know not who I am. Let those that have ears, hear." Quiet the mind, be free of clinging to molds and models and thought forms. Open the heart. Consume the emotions into the flow, the flow of all forms of life, until you are just flowing in and out.

As you get more disciplined, you keep the energy moving towards that point where form and formless meet. Were you to stay in the formless, your body which you left behind would disintegrate for there would be no consciousness to keep it going. There are all gradations, and some beings are ninety-nine percent in that ocean of formlessness and leave behind just a thread in form. There was a being walled up in a cave for twenty years, whom every year devotees would go to see and have darshan with what was a skeleton except the hair and the nails kept growing. He just left a thread behind to give darshan to the devotees.

Krishna, Christ, Durga, Kali—all of them the same. The ocean made manifest in different forms. Different strokes for different folks. Each a form you need, if you need form.

Methods and More

We come together, representatives of many forms, many methods. All the way from Krishnamurti who says there is no method, to Krishna Consciousness, or Fundamental Christianity, which says, "Our way is the only way," and it has much form. Where we meet is in what is common to all of our forms. And what is common to all of our forms is not another form. What is common to all of our forms is choiceless awareness, is pure love, is flow and harmony in the universe, the absence of clinging, spaciousness. You can call this "Buddha Mind," you can call it "the Heart of Allah." You could call it "Christ consciousness." You could call it "Yahweh," or "G—d." I have involved myself with many forms. Methods of Vipassana meditation, to make me more mindful, to quiet my mind and to bring it to one-pointedness. Devotional practices, worshipping the feet of my Guru and singing Hare Krishna and Sri Ram Jai Ram. Of Zen meditation, confronting a koan, or just sitting. Study, of the *Bhagavad Gita*, of *The Tibetan Book of the Great Liberation*, of Chung Tzu and Lao Tzu. Of the *I Ching* and the *Tao Teh Ching*, of the *New Testament* and the *Old Testament*, and on and on. How does it all come together? There is no form that represents the amalgam of all those things.

If you follow all these methods to the apex, you are pushed beyond form. You are pushed into the moment. The merging with God is right here. Be right here, aware of sitting here, aware of the self-definition that you're creating in your own mind. Aware of your ears listening. Aware of me speaking. Aware of the traffic outside. Aware of the feelings in your body. Aware of your mind grabbing at this and that. Just sit with me in this awareness. There is nothing we have to do, just come into this moment. Don't collect it, don't judge it. Just bring in more awareness. Watch your mind. Listen to yourself. Feel your heart. Is it flowing? Breathe in and out of the middle of your chest, as if there were a flow moving in and out of your heart with every breath. Flowing. Present. Here. More here. More. Let go of your expectations a little more. Of your definitions of who you are, of what God is. Of where you're going, of where you've come from. Of your emotions: sadness, happiness. Don't push them away. Notice them. Acknowledge them. Give them space. They are all part of the flow. Your senses, your memories, your plans, your models; all of it. Passing show. Forms being created, existing, and disappearing back into formlessness. Here in the moment. Right here. For the end result of everything that you and I have been sharing for years and years is not there or then or "maybe," or "perhaps," or "if only," or "as soon as I . . . ." This is it. Look at the stuff in you that's keeping you from being here at this moment. Judging. Waiting. Trying to experience. "I can't get it. I still feel separate." That thought—there's the problem right there. Let it go. The quiet mind. Choiceless awareness. Perfect flow and harmony. No you. No self-consciousness. Not, "I am trying to become enlightened." In meditation, there is no meditator. Meditation just *is*. Meditation is the act of openness. Of spaciousness. Of presence. Of is-ness.

So why are we joining all these clubs? Why are we paying all these heavy dues? What is it all about? Are all methods to be avoided? It doesn't seem so. But it does seem useful to see them in perspective. Methods are the ship crossing the ocean of existence. If you're halfway across the ocean, it's a little silly to decide methods are unnecessary if you don't know how to swim. But once you get to the far shore, it would be useless to keep carrying the boat. The game seems very simple: Methods are not the thing itself; methods are traps. You entrap yourself in order to burn out things in you which keep you from being free. And ultimately the methods spew you out at the other end, and the method disintegrates into nothingness. Every method: the Guru, chanting, study, meditation, practices; all of it. For the end result is "nothing special."

If we take knowing God as being always in meditation as you act all day long, as choiceless awareness, as being clear with no judgments, no opinions, no clinging, no pushing and pulling, no this nor that— we will experience what it means to know and be in God.

But if there are any experiences that you crave other than being free of the separation between experiencer and experience, that's what you need to concern yourself with. Not with fear about your cravings, but bringing to them consciousness and truth and quietness. For every teacher, every life experience, everything you notice in the universe is but a reflection of your attachments. That's just the way it works. If there is nothing you want, there is nothing that clings. You go through life free, collecting nothing. When you collect a sight, or collect a picture, or a record, or a stereo, or a relationship, or a teacher, or methods, it's just more clinging. Use them all, be with them, enjoy them, live fully in life; but don't cling. Flow through it, be with it, let it go. As you quiet and listen to hear how it all is, then you will relate to all of it in a harmonious way, in a way in which there is not exploitation. Harmonious in the way you relate to the floor you're sitting on, to the person next to you, to the night air, to the world you have to live in.

If I can hear it, right where I am, whatever space I might be in that moment, when there is no clinging, when I am neither attached to emptiness or form, I am free. If I push away the physical existence in order to get into "a space," if I'm only comfortable when I'm hanging out with Krishna and I can't stand my mother-in-law, I'm trapped. No clinging anywhere. And then the moment gets so rich, right here is it all. Every astral plane, physical plane, every level of consciousness, every mental state, all the emptiness: all of it, right here. Just a quiet mind hears it all.

It's your purity that calls forth the teachings, it's your acknowledgement of who you are. It's your quietness, it's your just opening yourself to the space you exist in. Instead of judging and pushing and pulling; opening to it. Just consuming the stuff, letting it all flow through you and in and out of you; just allowing it to pass.

If your method is Vipassana meditation, you're just noticing everything in the universe around you with bare attention. Maybe starting with the simple thing of noticing the breath go in and out of your nostril, or go up and down in your solar plexus. The things you'll have to let go of are self-pity, feelings of unworthiness, feelings of inadequacy, clinging to a judging mind, attachments to desires which see things as objects, which push the universe away. There is simply awareness noticing each element of the mind-body process as it comes and goes, but "nobody" watching.

If your method is the Guru, then you look at the Guru and the Guru keeps changing before your eyes. First you've got this form and then this form falls away, and then that form falls away. It's like Chinese puzzle boxes. You keep opening them and there's more inside. And you keep going until you realize you're just looking at a mirror. And all you're doing is cleaning, you're peeling yourself like an onion. And as you get purer and see more of your Guru. Till finally it's just one mirror looking at another, and no dust anywhere. Then there's no mirror. There's nothing. Your Guru disappeared into your own enlightenment. You ate your Guru alive. You and the Guru became one in God. That's the way the game works. That's the method of the Guru.

You should be open to all teachers and all teachings, and listen with your heart. With some you will feel you have no business. Others will pull you. Start to trust yourself. You have everything in you that Buddha has, that Christ has—you've got it all. But only when you start to acknowledge it is it going to get interesting. Your problem is you're afraid to acknowledge your own beauty. You're too busy holding on to your unworthiness. You'd rather be a schnook sitting before some great man. That fits in more with who you think you are. Well, enough already. I sit before you and I look and I see your beauty, even if you don't.

Do you realize, historically, how rare it is that this kind of a dialogue has existed with this much consciousness in it? Without everybody having eaten rye bread that went bad or something and getting ergot poisoning?

Once you find your lineage—and you can't go looking for it, you will be drawn to it, and it may not be in the form of a single teacher—it will merely be a way in which you view the universe. And through surrender into this lineage every act you do will be done from a space of greater clarity, will be an act not determined by your personal desires, but by the dharmic moment. It will be pulled forth from you. Just as these words are pulled forth from me by you. I have no identification with them, this book is just the transcript of words spoken to us listening, demanding that they be spoken. So whose book is it? When beautiful music is played on a violin, would you go up and thank the violin? I'm just the mouthpiece for a process. What you're doing when you read this book is you're touching your self. Forget me, I am passing show. You're touching yourself. Sooner or later you're going to have to acknowledge your beauty. But that acknowledgement isn't the end point. That's merely to override the acknowledgement of your ugliness, to which you've been clinging. Then both of them are going to have to go. For the end point isn't self-conscious, sitting around saying, "Look how beautiful I am." The end point is just being in the present moment.

And when you finish with your lineage and you get spewed out the other end, then you will look around and you will see that all methods get to the top of the mountain. And that you can find God in everyone. Then you no longer are a Buddhist or a

Hindu or a Christian or a Jew or a Moslem. You are love, you are truth. And love and truth have no form. They flow into forms. But the word is never the same as that which the word connotes. The word "God" is not God, the word "Mother" is not Mother, the word "Self" is not Self, the word "moment" is not the moment. All of these words are empty. We're playing at the level of intellect. Feeding that thing in us that keeps wanting to understand. And here we are, all the words we've said are gone. Where did they go? Do you remember them all? Empty, empty. If you heard them, you are at this moment empty. You're ready for the next word. And the word will go through you. You don't have to know anything; that's what's so funny about it. You get so simple. You're empty. You know nothing. You simply are wisdom. Not becoming anything, just being everything.

All of the things I would share with you are unspeakable. It is only now that we have this book out of the way that we can start to dance into the realms where we look into one another's eyes and know what is not knowable. We are that thing. For ultimately you will transcend knowledge. And you will be wise, a simplicity out of which comes the wisdom of your being. A human birth is a very precious matter. You have all the ingredients necessary to know God fully in this lifetime. That we all reached forth to meet here is itself incredible grace.

"Rejoice in the Lord always and again I say rejoice.
Rejoice rejoice and again I say rejoice.
Rejoice rejoice and again I say rejoice."

ABOUT THE AUTHOR

RAM DASS, under his original name, Richard Alpert, first became widely known in the early sixties, when the public became aware of the experiments he and his colleague, Timothy Leary, were conducting with psychedelic drugs at Harvard University. Richard Alpert was born in Boston in 1931. He received his Ph.D. in psychology from Stanford University, then taught at Stanford, the University of California at Berkeley, and finally at Harvard until 1963. While at Harvard, he taught and researched in the fields of human motivation, Freudian theories of early social development, cognition, and clinical pathology, and served as a psychotherapist with the Harvard University Health Services. In 1961 he had his first personal experience with the effects of psychedelic drugs. The experience was a profound and unsettling one. Following it, he joined with Timothy Leary and others in a research program at Harvard concerning the altered states of consciousness created by psychedelic drugs such as LSD. In the course of many experiments, Alpert took over three hundred dosages of psychedelic drugs. Applying his training as a psychologist, as well as his experience of five years in psychoanalysis, he observed certain shifts in his own psychodynamics. He also became profoundly aware of the limitations of the drug-induced experience. In 1967 he went to India where he knew that there were spiritual traditions that dealt with occurrences which paralleled those he experienced with psychedelics. After months of searching, Alpert found his guru, Neem Karoli Baba ("Maharaj-ji"), a man who had realized in his own life the wisdom that Alpert had only glimpsed under psychedelics. Maharaj-ji named him Ram Dass and directed him to study Raja Yoga. After several months, Ram Dass returned to the United States, and through lectures and eventually the books *Be Here Now, Journey of Awakening: A Meditator's Guidebook* and *Grist for the Mill*, he began sharing what he had learned. Since his first visit, Ram Dass has returned to India and the Far East from time to time to continue his studies.

LOOKING IN

Books that explore Eastern and Western spirituality and examine man's concepts of God and the universe.

☐	11710	THE GOSPEL ACCORDING TO PEANUTS Robert L. Short	$1.50
☐	12596	"WITH GOD ALL THINGS ARE POSSIBLE!" Life Study Fellowship	$1.95
☐	12009	THE GREATEST SALESMAN IN THE WORLD Og Mandino	$1.95
☐	12307	MYTHS TO LIVE BY Joseph Campbell	$2.50
☐	10176	THE BIBLE AS HISTORY: A CONFIRMATION OF THE BOOK OF BOOKS Werner Keller	$2.50
☐	12491	THE PASSOVER PLOT Dr. Hugh J. Schonfield	$2.25
☐	12315	I CHING: BOOK OF CHANGES Ch'u Chai, ed. with Windberg Chai	$2.25
☐	11578	JOURNEY OF AWAKENING Ram Dass	$2.95
☐	11475	WHATEVER BECAME OF SIN? Dr. Karl Menninger	$2.25

INTIMATE REFLECTIONS

Thoughts, ideas, and perceptions of life as it is.